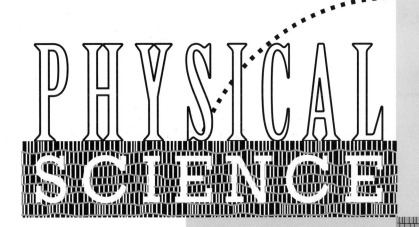

PHYSICAL SCIENCE

DISCOVERING MATTER & ENERGY

Nabil Kozman, M. Ed.
Lindsay Place High School
Lakeshore School Board
Pointe Claire, Québec

Barbara De Lorenzi, Ph. D.
John Abbott College
Ste-Anne-de-Bellevue, Québec

Shahid Jalil, M.Sc., M.Ed.
John Abbott College
Ste-Anne-de-Bellevue, Québec

Brenda Shapiro, Ph.D. (1943–1990)
Winston Churchill Collegiate Institute
Scarborough Board of Education
Scarborough, Ontario

Stan Shapiro, Ph.D.
Dr. Norman Bethune Collegiate Institute
Scarborough Board of Education
Scarborough, Ontario

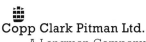

Copp Clark Pitman Ltd.
A Longman Company
Toronto

ISBN 0-7730-5041-8

Editorial development: Michael Webb
Editing: Susan Marshall
Design: Michael van Elsen Design Inc.
Cover design: Liz Nyman
Cover photograph: Barbara Stephens
Technical art: Allan Moon and Dave McKay
Creative art: William Kimber, William Laughton, and
  Kyle Gell
Photo research: Susan Marshall
Typesetting: Compeer Typographic Services Ltd.
Printing and binding: Arcata Graphics

Canadian Cataloguing in Publication Data
Main entry under title:

Physical science: discovering matter and energy

Includes index.
ISBN 0-7730-5041-8
1. Chemistry.   2. Physics.   I. Kozman, Nabil.

QD33.P59 1991   540   C91-094771-6

Copp Clark Pitman Ltd.
2775 Matheson Blvd. East
Mississauga, Ontario
L4W 4P7

Printed and bound in the United States of America.

2 3 4 5   5041-8   95 94 93 92

Reviewing Consultants

Ken Elliott
Protestant School Board of Greater Montréal
Montréal, Québec

George Ladd
Lakeshore School Board
Beaconsfield, Québec

Darlene McCrae
Queen of Angels Academy
Dorval, Québec

Contents

PHOTO CREDITS

This book is different from any science text that you have used before. You, the student, may think of science as a collection of facts to be memorized, and of science experiments as activities you carry out to confirm what you already know. Real science is not like that. This book is an attempt to give you some learning experiences that more closely resemble what scientists do.

Any scientist doing research has an idea of what the outcome of an experiment *might* be. But scientists are often wrong. The most interesting experimental results are often the least expected ones. This course will place you in many situations where you will not be sure of what to expect. Use your best guess and see how it compares with what you actually observe. Do not be afraid to make mistakes (as long as they do not involve a safety hazard). Above all, *learn* from what you see, and use new information as building blocks for further learning.

Science does not exist in isolation. It is connected with a vast range of social issues, from environmental protection to standard of living. This book will encourage you to think about the issues that surround science and to form your own opinions about how various issues should be handled.

If this book convinces you that science is a part of your everyday life and that the skills you learn in science help you to make informed decisions, then it will have succeeded in what it set out to do. We, the authors, hope that you enjoy this book and that you profit from using it.

1 SETTING THE SCENE

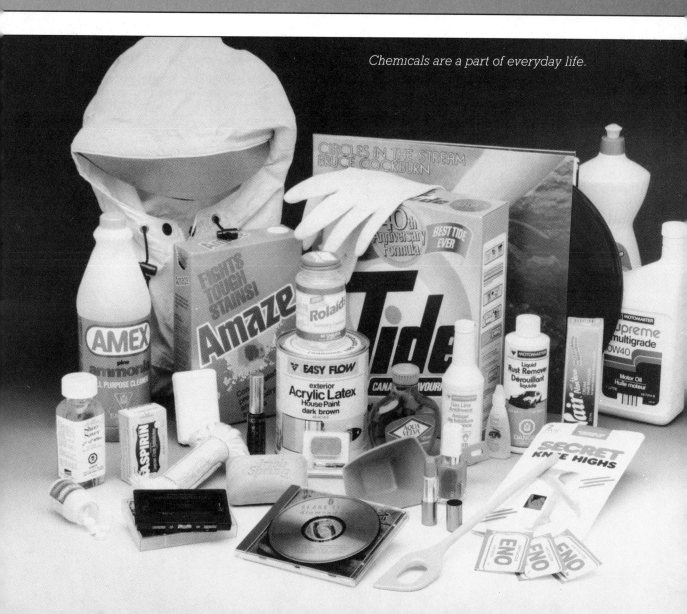

Chemicals are a part of everyday life.

CONTENTS

Fig. 1 *Where are the chemicals in these rooms?*

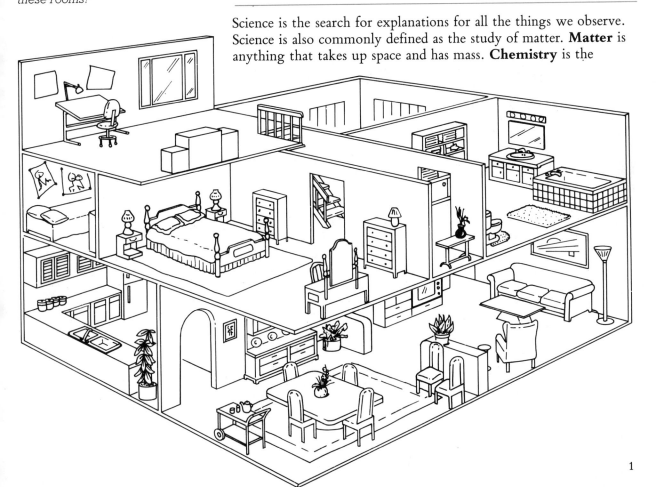

1.1 You're Surrounded

Science is the search for explanations for all the things we observe. Science is also commonly defined as the study of matter. **Matter** is anything that takes up space and has mass. **Chemistry** is the

1

branch of science that tries to explain the structure of matter and the changes matter undergoes. Can you think of something that does *not* take up space or does *not* have mass? Make a list of things that are not matter and share this list with your classmates.

Now let's turn to a group activity. Form your group with three other students. Pick a room in the house in Figure 1 that no other group has chosen.

List all the items you can see in your chosen room that are made up of chemicals. Use felt markers to write your list on a large piece of paper. When you have written everything you can think of, post your list in the classroom.

Study the lists produced by all the groups. In the rooms, is there anything that is not made up of chemicals?

You now know that you are surrounded by chemicals. But what about your own body? Is it made up of chemicals? In fact, the human body is a chemical factory second to none. All the time chemical reactions are taking place in the cells of our bodies. The foods that we eat are also made up of chemicals. The idea that naturally grown vegetables and fruits have no chemicals is not true. What they don't have are contaminants such as insecticides and fungicides.

1.2 Technology Brings Science into Our Homes

Many of the things that make our homes more comfortable and our everyday lives more enjoyable are the result of applying scientific knowledge to solve very practical problems. This process is called **technology**.

Think of the glass of milk on your kitchen table. Buying milk from the store is certainly more convenient than milking the family cow. To produce that glass of milk, the farmer used a milking machine to milk the cows. The milk from the entire dairy herd was then pumped into a refrigerated tanker truck for transportation to a plant, where it was pasteurized to kill any harmful bacteria. The milk was also homogenized to prevent it from separating into layers of cream and water. Once it was put into waxed cardboard containers or plastic bags, the milk was distributed to supermarkets, where it was displayed in refrigerated shelves. From this description, how many technological advances can you list that helped to produce the glass of milk? Discuss your list with your classmates.

It is important to realize the distinction between science and technology. For example, a study of the quantity of heat released by a known quantity of burning fuel is an interesting science activity. Using your scientific knowledge of burning fuels and the properties of building materials to design a fireplace for your home is an example of technology.

The drive to develop new technologies often comes from scientific discoveries. To return to the above example of milk processing and packaging, you would only think of building a refrigerator once you know that food spoils more slowly when it is cooled.

Questions

1. There are many more products of technology in your everyday life. List ten things that you have done today that depend on technological advances.

2. For each activity in question 1, indicate each of the following:
   (a) a technological advance that makes the item possible
   (b) a scientific discovery on which the technology depends

1.3 Your Own Research Project

Research is an important part of science. Scientists are constantly looking for new information and for new ways of solving problems.

Before starting an experiment, a scientist must select a topic to study and then find out what is already known about that topic. For your own research project, you are going to pick a topic that interests you and find out everything you can about that topic.

Here are the ground rules:

1. Your topic must concern the properties, structure, or classification of matter.

2. The topic must relate to some aspect of your life.

3. You must show how the scientific, technological, and social aspects of this topic are interrelated.

Here's an example of how you might choose a topic. Suppose you are interested in sound. We generate sounds in all kinds of ways using different types or classes of matter, so the topic of sound obeys ground rule 1. You must now choose an aspect of your life affected by sound. You might think of entertainment and communications, or some environmental concern such as noise pollution. Let's say you are more interested in communications. Now you have satisfied ground rule 2. Perhaps you feel comfortable about showing how the scientific, technological, and social aspects of *all* communications are interrelated (ground rule 3). But perhaps not. You may think that this topic is too big and that you should look at just one type of communication. Since you are fascinated by the telephone, you decide to research the scientific, technological, and social aspects of the telephone. Now, you have a topic!

Come up with a list of four or five possible research topics of your own. Discuss them in class and choose the topic you like best. Make sure that your teacher approves your choice. If you have difficulty in finding a research project, ask your teacher for help.

Once you have your topic, there are many books, magazines, films, and software that you could use to research your topic. Are there people in your community whom you could interview? Are there factories, laboratories, or businesses in your community that you might visit in order to learn more about your topic? Also, think about ways of presenting the results of your research project. You need not limit yourself to a written report. Perhaps you could take photographs, make a video or audio recording, or write and act out a skit. Discuss all of these options with your teacher, who may decide to give you a checklist to help you plan and carry out the project.

1.4 Don't Live to Regret Something

Choosing a topic in section 1.3 is only the first stage in scientific research. Once you have defined an area of interest, found a specific topic, and learned what is already known about it, you will then want to learn something new about the topic. So you must design and carry out experiments to learn first-hand how different types of matter behave in various situations. Before you do this, it is essential that you know how to work safely in the laboratory. Remember, a laboratory accident may have serious consequences.

To help you understand how important it is to avoid laboratory accidents, try the following activity. First, choose *one* of these three situations:

- *Pretend that you are blind.* Wear a blindfold. Another student will watch over you so you can avoid hurting yourself.

- *Pretend that you have lost the use of your writing hand and the thumb and forefinger of your other hand.* Tie your writing hand behind you, and tape together the thumb and forefinger of your other hand.

- *Pretend that you have been badly hurt and have lost all your fingers.* Tape the fingers of each hand to the palm.

Once you are prepared, you are then ready to do the activity:

1. Set up a retort stand and fix a small ring clamp to the stand.

2. Place a funnel in the ring clamp so that the end of the funnel is inside a 250-mL flask and about 2 cm below the rim of the flask.

3. Pour 50 mL of water from a graduated cylinder through the funnel into the flask.

4. Pour 5 mL of water from the flask into a 10-mL graduated cylinder.

Write down the feelings that you experienced while doing the above activity. They may include a little frustration. Then, in groups of four, discuss how an accident could affect your lives.

Draw your image of the word "safety" on a large piece of paper and put the drawing up in the classroom. Alternatively, perform a charade on the word "safety."

1.5 Safety First

The following are some important safety rules for working in the laboratory. As long as instructions and safety precautions are followed, the laboratory does not have to be a dangerous place.

- *Always read the safety instructions before you start any part of an experiment.* This will reduce the chance of accidents.

- *Wear goggles* throughout activities that involve glassware or liquid chemicals (Figure 2).

- *Tie back long hair and do not wear loose clothing* (Figure 3).

Fig. 2 *Goggles protect your eyes from shattered glass and splashed chemicals.*

Fig. 3 *Long hair or loose clothing may catch fire in a Bunsen flame. What precautions are these students taking?*

- *Always wear shoes* (but not open-toed ones) in the laboratory. Some chemicals may have been spilled on the floor.

- *Always stand while doing an experiment.* In case of accident, you can get away faster if you are standing.

- *Never sit on laboratory benches.* Some chemicals may have been spilled on them.

- *Learn the fire-exit route.* In a fire, you need to get to the exit quickly and safely.

- *Learn the location of the fire alarm* nearest to the laboratory. You should warn others immediately of fire. Never pull a fire alarm as a joke.

- *Learn the location of the fire blanket.* If your clothes catch fire, you will need a fire blanket quickly. Wrap yourself in the blanket and roll on the floor.

- *Use tongs to handle hot equipment.* Remember hot metal and glass look like cold metal and glass.

- If your Bunsen burner does not light as soon as you turn on the gas and use a spark, *turn off the gas.* Wait a few minutes for the gas to clear and try again.

  When the gas is turned on, it comes out of the Bunsen burner. The longer the burner remains unlit, the more gas surrounds you. When it eventually lights up, you may be burned!

- *Never leave a lighted Bunsen burner unattended.* Someone may get burned because flames are not always easy to see.

- *Use a hotplate when heating flammable substances* (Figure 4).

- *Always clean your own equipment.* Although equipment may look clean, it usually is not!

- *To pick up chemicals, use a scoop or spatula that you have cleaned and dried.* Do not touch chemicals with your hands. They may hurt your skin.

- *After using a spatula with one chemical, always clean and dry the spatula before using it with another chemical.* If you do not clean the spatula, the second chemical will become contaminated with the first. The chemicals may react violently with each other.

- *If a chemical is stored in a container with a lid, you should always replace the lid properly after use.* Do not leave the container open for the second person. Replacing the lid avoids contamination of one

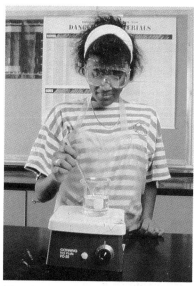

Fig. 4 *Use a hotplate when you heat flammable substances. Using the Bunsen burner may cause a fire.*

Fig. 5 *To detect the odour of a chemical, wave the fumes toward you. Do this only if your teacher asks you to. Never put the test tube under your nose.*

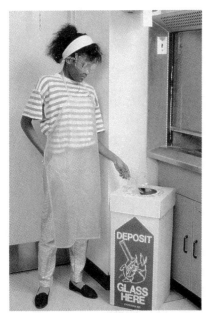

Fig. 6 *Dispose of your garbage properly. Don't put broken glass in the regular garbage containers; otherwise the people emptying them may be hurt.*

chemical by another. It also prevents reaction of the chemical with air and water vapour.

- *Use a clean glass rod when pouring a liquid from one container into another.* This prevents splashing of the liquid.

- *If you spill chemicals on your hands or clothes, rinse the chemicals off with water* (unless the instructions tell you not to use water). Water dilutes the chemicals and washes them off your skin or clothes.

- *Spilled chemicals should be cleaned up immediately.* Other people could get hurt by touching them.

- *Never try to push glass tubing though a hole in a stopper.* If the glass breaks, you may cut yourself. Ask your teacher for assistance.

- *Do not add water to a chemical.* Always add the chemical to water. Many chemicals, especially concentrated acids, will react violently if water is added to them. Adding them to water reduces the violence of the reactions.

- *To smell a chemical, use your hand to wave some of the vapour toward your nose.* Partly fill your lungs with air and take a *short* sniff of the vapour (Figure 5).

- *Never taste a chemical* unless the instructions definitely tell you to. If you are required to taste a chemical, it should not be contained in equipment that is regularly used in the laboratory because it may be contaminated with poisonous chemicals. Use special or new equipment.

- *When using electrical equipment, make sure your hands are dry and there are no exposed wires.* Otherwise you may get an electric shock.

- *Notify the teacher immediately if there has been an accident, or if any equipment is broken.*

- *If you burn yourself, immediately put the burned region under cold, running water.* Then notify your teacher. The cold water takes the heat away from the burn. Do not use ointments, which may cause problems.

- *Put any broken glass in the special "broken glass" container,* not in the regular paper garbage container (Figure 6).

- *Do not carry out any experimental procedure you devise yourself until your teacher has checked and approved it.*

- *Do not build an electrical circuit when any part of it is "live," that is, connected to a power supply.*

Symbols for Chemical Hazards

In this text, any source of danger from a chemical is identified by a symbol. The meaning of each symbol is explained below. Learn the meaning of these symbols. You will see the same ones on products that you buy.

Type of Hazard	*Symbol*	*Description*
Toxic or Poisonous		A substance that could be harmful to human health, could cause cancer or birth defects, or could contaminate, harm, or kill fish and wildlife.
Corrosive		An acidic or basic substance that could corrode storage containers or damage human tissue if touched.
Reactive or Explosive		A substance that could change rapidly (or even explode) if exposed to heat, shock, air, or water.
Flammable		A substance that could explode, catch on fire, or give off poisonous fumes or gases.

Symbolic Code for Degree of Hazard

Level 1 Caution (mildly hazardous)				
Level 2 Warning (moderately hazardous)				
Level 3 Danger (severely hazardous)				

To remind the class of the importance of laboratory safety, carry out the following:

1. Design a checklist or questionnaire on the safe operation of chemistry and physics laboratories in your school.

2. Design a campaign to promote safer procedures in a laboratory. This can be done in the form of a poster, a skit, a video, or a taped announcement.

1.6 How Safe Are You?

Check your knowledge of the correct safety procedures and use of equipment by completing the crossword puzzle in Figure 7. Your teacher will give you a large puzzle to fill in. (Do not write on the puzzle in the textbook.)

Fig. 7 *The safety crossword puzzle.*

Across

2. Heating equipment for flammable liquids
7. Worn at all times when doing an experiment
8. A flammable liquid that makes people drunk
9. A narrow piece of glassware that should never be more than one-third full when being heated
10. Equipment that must always be cleaned and dried before being used to transfer solids
13. A heating device used in the laboratory that should never be left unattended
15. A cone of _____ used to line a funnel
19. A piece of iron equipment attached to a retort stand to support a gauze or hold a funnel
20. _____ fool around in the laboratory
21. A box-like heating container used to heat materials over long periods
22. Dangerous to _____ glass tubing into a rubber stopper
24. A long piece of glass used to stir chemicals or to help when pouring liquids
25. Type of equipment which, if broken, must not be thrown in the regular garbage container
26. Equipment used to handle hot liquids in beakers

Down

1. Poisonous silvery liquid found in some thermometers
3. A device for measuring temperature
4. Cylinder to measure the volume of liquid
5. A device used to separate a liquid from an undissolved solid
6. A piece of glassware, with its top much narrower than its base, used to hold liquids
7. Metal mesh used to prevent glassware from cracking when heated with a Bunsen burner
8. A type of chemical which must be respected and handled carefully in the home and laboratory
11. A metal rod to which you attach clamps and other equipment
12. Fuel for the Bunsen burner
13. A piece of short glassware with no lid used to hold liquids and solids
14. A very expensive instrument used to weigh things (which must not be hot!)
16. A piece of glassware used when filtering a mixture
17. The way your laboratory equipment and work station should always be kept
18. A metallic piece of equipment used to attach other equipment to a retort stand
23. Metal instrument used to handle very hot items or to hold items in the Bunsen flame
24. For a series of weighings, use the _____ balance

1.7 Observations, Inferences, Hypotheses, and Theories

In science, we study many changes that occur around us. We try to understand these changes by performing experiments. During an experiment, we record what we observe. Then we try to explain what we have just observed.

Observations are the most important part of any scientific investigation, so it is important that you make and record your observations as carefully as you can.

To make observations you must use one or more of your senses.

Detecting	Sense Used
shape, colour, shininess	sight
odour	smell
texture	touch
sound	hearing
taste (sweet, sour, salty, bitter)	taste

If you use your senses only, without the aid of any instruments, to make observations, the observations are **qualitative observations**. For example: ''The watermelon has red flesh.''

Sometimes you make observations using instruments, such as balances, graduated cylinders, thermometers, and rulers. Then the observations are **quantitative observations**. They are measured quantities, written as numbers and usually followed by units. For example: ''The watermelon has a mass of 4.00 kg.''

Fig. 8 *What qualitative and quantitative observations can you make?*

Fig. 9 *Observation or inference?*

If a guess is added to an observation, the statement is known as an **inference**. The inference may or may not be correct, depending on the correctness of the guess.

An example of an inference is: "The watermelon has thick skin to protect it from insect attack." The observation is simply that the watermelon has a thick skin. By adding that the skin protects it from insect attack — which is just a guess — we have turned the statement into an inference.

A **hypothesis** is a suggested explanation for a number of observations. Suppose that three bulls are attracted by a fluttering

11

red rag. Your hypothesis might be that the colour red attracts them. If you predict that the fluttering red rag will attract another bull, and it does, then the idea that the colour red attracts them becomes a **theory**. In other words, a theory is a hypothesis that allows you to predict correctly something that you have not already observed. Theories are assumed to be correct until proven otherwise.

There is nothing to say that a theory has to be correct. For example, bulls cannot actually see the colour red! Their eyes see everything as shades of gray. So a more reasonable theory might be that the fluttering motion of the rag attracts them. How could you test this theory?

Even widely accepted theories may be proved wrong. An example is the theory held by the ancient Greeks that the sun and moon move around the earth. This theory was accepted for over a thousand years.

If every week you observe a person buying cat litter, you would probably develop a hypothesis that the person has a cat. If your hypothesis was a good one, you might predict that the person would accept some free cat food samples. If the person were to say that cat food was of no use and turn down the free samples, you would need a new hypothesis that matches the old observations and includes the latest piece of information or evidence. What would be your new and improved hypothesis? How would you develop your hypothesis into a theory? How could you test the theory?

Questions

1. Decide whether each of the following is an inference or an observation. Write in your notebook the letters "a" to "m". Beside each letter write "I" (for inference) or "O" (for observation).
   (a) When the candle burns, carbon dioxide is given off.
   (b) The candle is 10 cm long.
   (c) The candle has a diameter of 2.5 cm and is cylindrical.
   (d) A candle flame has three different colours.
   (e) The black smoke coming from the candle flame is carbon.
   (f) If oxygen is removed from around a burning candle, the flame is extinguished.
   (g) The length of the candle changes slowly as it burns.
   (h) The wick extends from the top of the candle to the bottom.
   (i) A burning candle makes no sound.
   (j) Carbon causes the yellow colour in the candle flame.
   (k) The candle flame has sharp sides and a ragged top.
   (l) In the candle flame, the wick is black except for the very tip, which glows red.
   (m) The candle flame gives off heat and light.

2. The English physiologist T. H. Huxley once wrote, "That is the great tragedy of science — the slaying of a beautiful hypothesis by a single ugly fact." Discuss what you think this statement means.

1.8 Taking a Good Look Around You

A visitor has just arrived from planet Vulcan, which is in a distant galaxy. Welcome to our home! No doubt the visitor finds it quite different from Vulcan. Since you are a science student, it is up to you to assist the space traveller in examining and recording information about substances that belong to our environment.

Bring to class at least six different types of substances typical of our environment that you think would interest our visitor. They could be naturally occurring or manufactured substances.

In order to choose six types of substances from among the many possibilities around you, it may help you to group the substances in some way. For example, you might decide that all plastics form one group. To group things, you have to be aware of how they are alike and how they are different. Sorting things in this way is known as **classifying**.

When scientists classify matter, they look for ways of dividing a large group of different substances into smaller groups each with a common characteristic or **property**. One such simple test is determining the physical state of the substance at room temperature and at a specific pressure. Under these conditions some substances are gases, others are liquids, and still others are solids. What are some of the properties that distinguish gases, liquids, and solids from each other?

When you have collected all the items you want to show our visitor and have shown them to her, then answer the following questions posed by the Vulcan:

1. What do all these substances have in common?

2. What sort of measurements could be made on all the substances selected?

3. How could we divide them into smaller groups for easier study?

1.9 Classification

If we look around in our day-to-day environment, we come across all kinds of different substances, such as water, milk, sugar, vinegar, hair sprays, toothpaste, mouthwash, and so on. There are millions and millions of different substances known to us, and every day a few more are discovered. The first thing a scientist wants to do as soon as a new substance is discovered is to observe it closely and compare it to other known substances, looking for similarities and differences. The scientist will be looking at other characteristics besides the substance's physical state.

When we classify substances, we group together those that have common characteristics. At home you might classify the audio tape cassettes in your collection according to the performer. How else might you choose to classify the cassettes in your collection? If you have a collection of seashells or of rocks, how might you classify the items in your collection?

Fig. 10 *Classifying is important in daily life.*

| Investigation | *1.10 Properties Of Matter* |

In this investigation, we will examinine various kinds of matter. As you already know, just looking at them makes it possible to organize them into three major categories — solids, liquids, and gases.

Now we will explore the characteristics of each category. Perhaps we can subdivide each major category into smaller groups. Each smaller group would have similar characteristics.

Be sure to prepare a data chart to record all your observations. The chart is useful for comparing characteristics. You might want to use a chart such as the one below.

Name of substance	Test	Observation

property.

Materials

lab aprons

safety goggles	power supply
matches	conducting wire
wooden splints	light bulb
test tubes	electrodes
test tube rack	selection of gases
corks	selection of liquids
magnifying glass	selection of solids
watch glass	various indicators

Conductimeter

generate 3 gases

6

6 regular &

'how many 6 do I need'

5

7

hole stopper

Method

also refer to p 35 fig. 9.

1. *Glowing Splint Test:* Light a wooden splint with a match and blow it out gently so that it only glows. Insert the glowing splint into each of the gaseous substances in a test tube. Observe what happens in each of the containers.

2. *Lighted Splint Test:* Light a wooden splint, but this time do not blow it out. Test each gas in a test tube by bringing the lighted splint to the mouth of the tube. Observe what happens in each case.

3. *Limewater Test:* Add several millilitres of limewater (calcium hydroxide solution) to a sample of each of the gaseous substances in a test tube. Shake up the covered container. Observe what happens to the limewater in each case.

4. *Indicator Paper Test:* Dip a piece of indicator paper containing cobalt chloride to each of the liquids in your collection. Observe the colour of the paper in each solution.

starting point

5. *Litmus Paper Test:* Follow the same procedure using indicator paper containing litmus. Observe any colour changes.

Be careful not to touch the electrodes while the electricity is turned on. You may get a rude shock!

6. *Testing Conductivity:* Dip the electrodes from a simple apparatus such as the one illustrated in Figure 11 into each of the liquids in your collection. Write down what you observe. Be sure to rinse the electrodes with distilled or deionized water after each dunking! Also test each of the solids to see if they will conduct electricity.

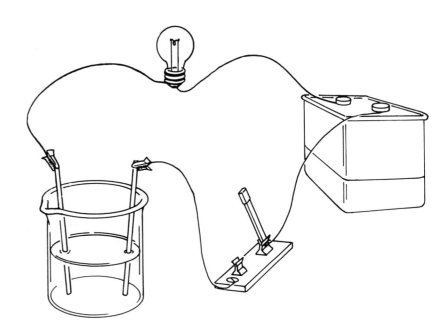

Fig. 11 *A simple conductivity apparatus.*

7. *Examining Solids with a Magnifying Glass:* Use a magnifying glass to examine the shape and colour of the particles in each of the solids. Describe what you observe under magnification.

8. *Ability to Dissolve in Water:* Put a sample of each solid into a different test tube. Add water to each test tube, cork the test tube, and shake up the contents to determine if the solid will dissolve. Record your observations in each case.

Be careful when using the acid and the base solutions in these tests. If you spill any or get any on your skin, wash it off immediately with lots of water. Ask your teacher for assistance.

9. *Hydrochloric Acid Test:* Put a sample of each of the solids in a different test tube. Add a few millilitres of hydrochloric acid solution to each sample. Observe what happens.

10. *Ammonia Solution Test:* To a sample of each of the solids, add a few millilitres of ammonia solution. Observe what happens.

Follow-up → *very important*

1. List all the substances in each category—solid, liquid, or gas— which behaved similarly in a given test.

2. For each test, is there any substance that behaves like no other substance? For example, think of the effect that oxygen had on a glowing splint. Did any other gas also behave in this way?

3. We refer to the effect that oxygen has on a glowing splint as a **characteristic property** of oxygen. Suggest a definition for the term ''characteristic property.''

4. Describe a characteristic property for each substance that you tested in this investigation.

5. We refer to properties observed for two or more substances in a group as ''non-characteristic'' properties. Why?

6. For each substance you examined, which of the properties are characteristic and which are non-characteristic?

7. How did you decide which properties are characteristic of a substance and which are non-characteristic?

8. Your observations in this investigation were all qualitative. Quantitative observations are also used to classify and distinguish substances. Use your knowledge from earlier science courses to describe some types of quantities you could have measured to tell substances apart. A common example is the boiling point. Give at least two other quantities.

9. Give at least two types of quantitative observations that are non-characteristic of a substance.

10. There is a saying: ''If it looks like a duck, walks like a duck and quacks like a duck, chances are it is a duck.'' Similarly, if the list of characteristic properties of an unknown substance matches the characteristic properties of a known substance down to the last item, then you can be reasonably, although not absolutely, sure the substances are the same. What is the substance described by this list of properties?

    It is a clear, colourless, and tasteless liquid.
    It freezes at 0°C and boils at 100°C.
    It dissolves a great many things including salt and sugar.
    Wood floats on its surface, but stones do not.
    Oil does not mix with it. Instead, oil floats as a layer on top of this liquid.
    It expands upon freezing.

    What do you think this substance is?

| *Investigation* | *1.11 Identifying a Sample* |

With all the excitement created by the arrival of our visitor from Vulcan, someone has forgotten to label a few samples left from investigation 1.10. Unmarked bottles can be a real problem in a laboratory. Why do you think this is true?

Now you have a mystery to solve. You will have to determine which of the samples you studied in investigation 1.10 corresponds to the substance in the unlabelled bottle.

Take a good look at the sample. Which of the major categories of substances does it come from? Which one of the substances in that group do you think it is? Write down your guess. How will you know if your guess is correct? What are the characteristic properties that substances in that group have in common? Which properties are characteristic of individual members of the group? Which tests would you use to identify your substance?

Materials

Make a list of the chemicals and equipment you will need to perform the test(s) you have chosen.

Method

1. Write down each step in your testing procedure.

2. Have both the list of materials and the procedure approved by your teacher before you begin.

Follow-up

1. Which of the possibilities did your sample turn out to be?

2. Which of the tests enabled you to decide its identity?

3. Write a brief report on the experiment you have just conducted. Explain how you figured out which tests to perform and how the results helped you identify the sample.

1.12 Consumer Goods and Properties

When you go to the store to buy juice or a soft drink, what type of beverage container are you likely to choose? Do you prefer to buy a beverage in a glass bottle? a plastic bottle? an aluminum can? a plasticized cardboard container? Which type of container would you prefer if the volume of liquid and the price were the same?

Fig. 12 *Some common types of food containers.*

When you make decisions of this kind, you are making a choice on the basis of the characteristic properties of a material.

Both glass and plastic bottles let you see what is inside. What are the advantages of plastic bottles? of glass bottles? What are the disadvantages of each type of bottle?

Aluminum cans are popular for soft drinks. Most vending machines are stocked with aluminum cans. Why do you suppose they are so popular? What are the disadvantages of aluminum cans?

Fruit drinks are often packaged in cardboard containers that are coated to make them waterproof. Light will destroy the vitamin C in fruit juices, making the shelf life of these juices much shorter if the containers are plastic or glass. A potential problem with aluminum cans is that the juices may react over time with the aluminum in the cans. This reaction will affect the taste of the juice. All types of container materials have advantages and disadvantages. The examples above illustrate ''consumer appeal'' and the containers' effects on the products they contain. But there are many other advantages and also disadvantages to consider, including costs and environmental effects.

Polyethylene is a type of plastic that is used in beverage bottles. This plastic is one of the by-products of the production of gasoline from petroleum—a process called ''cracking.'' Petroleum is a non-

CHEMICAL TIDBITS

Uniform substances are called "homogeneous."

THE MAKING OF GLASS

Glass making is one of the oldest technological processes. Objects made of glass were discovered in the tombs of ancient Egypt. Glass can be made by heating a mixture of sand (which is essentially silicon dioxide), sodium carbonate, and limestone (calcium carbonate). Once the mixture cools, it becomes hard, transparent, and uniform.

Different colours and different properties of glass can be produced by altering the proportions of the ingredients and by adding various metal oxides. If boron oxide replaces some silica, the resulting borosilicate glass is more heat resistant. Beakers and glass cookware are made from this kind of glass.

Fig. 13 *Glass is an important product.*

The raw materials used in the production of glass are not expensive and are plentiful, but considerable energy is needed to melt and shape glass. Although glass is breakable, it does not rot, corrode, or decompose in any way. Fortunately, glass can be melted down and re-formed into new objects, as long as different kinds of glass are separated before they are melted.

renewable resource, and plastic, like glass, does not disintegrate for very long periods of time. So energy is saved and our environment is protected if a plastic such as polyethylene is recycled. The technology required for the recycling of plastics has been developed. Scrap polyethylene has always been melted down and re-formed in the plants that produce beverage bottles. What do you think prevents more widespread recycling of plastic bottles?

Although aluminum is plentiful in Canada, it is a non-renewable resource. Aluminum is separated from an ore called bauxite. This mineral contains aluminum oxide, along with iron oxides, silicates, and other impurities. Large amounts of electricity must be passed through molten bauxite to obtain pure aluminum. About 18 700 kWh of electricity is used in the production of each tonne of aluminum. (By contrast, a 60-W light bulb uses just over 500 kWh if it is left on for a whole year.) Aluminum mills discharge a great deal of pollution into the air and metal oxides into waterways. Aluminum's resistance to corrosion is both its greatest advantage and its greatest disadvantage. It has been estimated that it would take over 500 years for a can made of aluminum to corrode completely. Why is the recycling of aluminum important both for the economy and for the environment?

As for paper-based containers, did you know that each year hundreds of millions of trees are harvested in Canada and that a seedling takes 65 years to grow to the height of the tree? that half of what we throw away is paper?

Next time you reach for a container of a soft drink, think about the consequences of your choice.

Questions

1. Select a type of consumer product that comes in a variety of forms or packaging. The products within each type must all have the same use. Possible choices include cookware, shingles, tiles, paving materials, fabrics, wheel rims, fuel, recorded music, wires for transmission of electricity or messages, soaps and detergents, photographic film, car bodies, and so on.
   (a) Within your chosen type, what are the advantages and disadvantages of each consumer product? Evaluate each product in terms of its intended use.
   (b) How are the various types of this consumer product manufactured? What raw materials are consumed? How much energy is involved in the production process? For the answers to these questions, consult reference materials.
   (c) What effect does the product and its production have on your province's economy and environment?

P O I N T S · T O · R E C A L L

- Chemistry is the branch of science that tries to explain matter and the changes it undergoes.
- Matter is anything that takes up space and has mass.
- There are many chemicals in and around you.
- When you work in the laboratory, you must work safely so as to avoid hurting yourself and others.
- Technology is the application of science to solve practical problems.
- Observations can be qualitative or quantitative.
- An inference is a statement that includes an observation and a guess.

- A hypothesis is a possible explanation of observations.
- A theory is a hypothesis that makes correct predictions.
- Scientists classify substances into groups with common characteristics.
- Matter can exist as a solid, liquid, or gas.
- A characteristic property is one that can be used to identify a substance.
- When planning a new product, you must consider energy use and the environment.

R E V I E W · Q U E S T I O N S

1. List five items found in your bedroom that are made up of chemicals.
2. List three items that you may find in the kitchen that contain colouring agents.
3. List three items that you may find in the bathroom that contain perfume.
4. Some people feel that non-active ingredients in products should be banned. What do you think? Explain your answer.
5. Why is safety so important in the chemistry laboratory?
6. Give at least three safety precautions you would use when boiling a liquid in a beaker.
7. What special precautions must you take when carrying out reactions that give poisonous products?
8. Give two reasons why you should never weigh hot objects on a balance.
9. Choose five types of chemical products and briefly indicate how they are used in your daily life.
10. Give two examples, besides those that are given in the textbook, that illustrate the distinction between science and technology.
11. Suppose you develop the hypothesis that the earth is flat.
    (a) What kind of evidence might suggest this hypothesis to you?
    (b) What kind of evidence might turn your hypothesis into a theory?
    (c) What kind of evidence would disprove this theory?

12. Why is it difficult to classify materials on the basis of appearance alone?
13. Classify the following substances according to their physical state at room temperature.
    (a) gasoline
    (b) salt
    (c) alcohol
    (d) oxygen
    (e) baking soda
    (f) helium
    (g) sugar
    (h) vinegar
    (i) carbon dioxide
14. List all the characteristic properties you can think of for the following substances.
    (a) iron
    (b) copper
    (c) graphite
    (d) milk
    (e) helium
15. What do you think is the most economical material to use for bottling soft drinks? Which is the most environmentally friendly material? Give reasons for the choice you make.

2 CHANGES IN MATTER

Fire is an indication that a chemical change is taking place.

CONTENTS

Fig. 1 *Learn how to keep fire out of your home and how to put out a fire.*

2.1 Our Changing Environment

Have you ever thought how boring your life would be if there were no change of season? There would be no anticipating the first snowfall, no planting of tulip bulbs that would burst through the earth in the spring, no looking forward to those long, warm days of summer, no leaves turning red and yellow. The falling leaves signal the beginning of yet another cycle of seasonal changes.

But there are other, smaller changes that are taking place around us all the time and that we take for granted. Some changes only make something *look different* like the melting of an ice cube or the shattering of glass. Such a change is temporary. You can re-freeze the water, and melt and re-form the piece of glass. Other changes make something that is *totally different*. Burning leaves is an example. Such changes are permanent. You cannot take what you are left with after a fire and turn it back into what you started with.

Look carefully at a safety match. Feel its shape and texture. Does it smell? Now strike the match and watch the match change. What do you smell? Look at the colours in the flame itself. Hold your open palm about 5 cm away from the burning match. What do you feel? Blow out the match and take a look at what remains. How has the match changed? If you strike the match again, will it catch fire? Is this change permanent or temporary?

CHEMICAL TIDBITS

Fig. 2 *Dust explosions are a hazard in grain elevators.*

FIRE SAFETY

Have you ever been advised by a fire safety officer on how to keep your home safe from fire (Figure 1)? If so, the officer might have told you that there are three requirements for a fire to occur:

- fuel
- oxygen
- a sufficiently high temperature

In groups of four, discuss the following questions:

1. How would you light a campfire or a fire in your fireplace? Write down each step and the reason for it.

2. Based on the requirements for a fire listed above, how would you put out a fire?

3. What would you do if you were in a building that caught fire?

4. Develop fire safety rules for your own home.

When you are finished, post a summary of your group's discussion. The class can then compare results.

The fire safety officer might also have told you not to accumulate rubbish, newspapers, and junk in your home or garage. Oil, paint, and polish on cleaning rags slowly react with oxygen. This reaction produces heat. The heat may build up sufficiently to set the rags on fire. This is an example of **spontaneous combustion**, or burning that starts by itself.

Burning can also occur much more quickly in an explosion or a flash of flame (Figure 2). If you have a cloud of powder, such as sawdust, grain dust, coal dust, or flour, a spark can set the whole cloud on fire almost at once. This is because the materials in powdered form have a large surface area exposed to the air, so they can react more easily with oxygen. If the powder catches fire in an enclosed space, like a grain elevator, an explosion may occur.

Questions

1. Look for objects in your environment that undergo changes. The changes may be either temporary or permanent. Bring the objects to class and discuss the changes you have observed with your classmates.
2. A permanent change produces something entirely new and different. List the objects that undergo permanent changes. What do you think causes each of these changes?
3. A temporary change can be reversed. For example, freezing water produces ice, melting ice produces water. List objects that undergo temporary changes. What do you think causes each of these changes?

| Investigation | *2.2 Types of Changes* |

As you have noticed in your daily life, matter undergoes many different changes. In this investigation, we will examine some chemicals and allow them to change in different ways. In each case, pay close attention to the substance before and after a change takes place and to the process of change itself. Follow the directions of your teacher and the rules for safety in the laboratory.

Materials

safety goggles
2 watch glasses
test tubes
test tube rack
retort stand
clamp
Bunsen burner
magnet
one 250-mL beaker
hammer
tweezers
iron filings
sulfur powder
magnesium ribbon
marble chips (calcium carbonate)
dilute hydrochloric acid
dilute sodium hydroxide solution
calcium chloride solution
sealed tube of iodine

Heat the test tube in the hottest part of the Bunsen burner flame, that is, at the tip of the inner cone of flame.

The fumes from burning sulfur are dangerous. Heat the mixture in a fume hood.

Do not forget your safety goggles!

Fig. 3

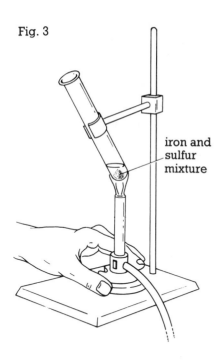

iron and sulfur mixture

Do not look directly at the reaction! You could damage your eyes. Watch the reaction through a piece of cobalt blue glass.

Method

I. *Mixing Iron and Sulfur*

1. Put a small sample of iron filings and one of sulfur powder on two separate watch glasses.

2. Examine the appearance and determine the magnetic properties of pure iron and of pure sulfur.

3. Mix them thoroughly and then try to separate them by using a magnet.

4. Put a mixture of iron and sulfur into a small test tube to a depth of about 1 cm. *In a fume hood*, heat the mixture until it glows (Figure 3).

5. Allow the test tube to cool and then remove the contents. (To shatter the test tube, wrap it first in a cloth and then gently hit it with a hammer. Use tweezers to remove the pieces of broken glass.) Examine the appearance and determine the magnetic properties of the contents.

(a) What do the iron filings look like before mixing and heating? What does the sulfur powder look like before mixing and heating?

(b) Which chemical is magnetic? How do you know?

(c) Which chemical is more crumbly or brittle? How do you know?

(d) What happened when you tried to separate the iron from the sulfur once the chemicals were mixed together? Suggest how iron-containing materials are separated from other metals at a junk yard.

(e) What happened when you heated the mixture of sulfur and iron? What did the contents of the test tube look like once they had cooled? Were you able to separate the iron from the sulfur in this material?

(f) Why do you think you were not able to separate the iron from sulfur after the substances had been heated?

II. *Burning Magnesium*

1. Cut off a 5-cm to 6-cm piece of magnesium ribbon.

2. Grasp it with tweezers and ignite it with a match or a Bunsen flame. (The magnesium may take a moment to catch fire.)

(a) Describe the appearance of the piece of magnesium ribbon. Is magnesium a hard or soft metal? What makes you think so?

(b) What happened to the magnesium when you heated it?

(c) What was the result of burning the magnesium? Does this result resemble the magnesium in any way? Is it soft and bendable like the magnesium, or is it brittle like the pure sulfur in part I of this investigation?

Fig. 4 *The reaction when a non-electronic flashbulb goes off is the same as when you heat magnesium.*

III. *Mixing Calcium Carbonate and Hydrochloric Acid*

1. Add one or two marble chips to a test tube containing a few millilitres of hydrochloric acid. (Marble chips are made of calcium carbonate.)

2. Observe what happens.

(a) What do the marble chips (calcium carbonate) look like?
(b) What happened when you put the marble chips into the hydrochloric acid?
(c) Has a change occurred? How do you know?

IV. *Mixing Calcium Chloride and Sodium Hydroxide Solutions*

1. In a test tube, mix three or four drops of sodium hydroxide solution with an equal volume of calcium chloride solution.

2. Observe what happens.

(a) What did each solution look like before mixing? What did the mixture look like?
(b) Has a change occurred? How do you know?

Handle all solutions with care.

The material that forms when the calcium chloride and sodium hydroxide solutions are mixed does not dissolve in water. It is called a **precipitate**.

Obtain a sample of iodine in a sealed tube. Iodine is quite toxic. If breathed or swallowed, it could make you seriously ill. Perform this part of the investigation in the fume hood.

The process in which a substance goes directly from the solid state to the gaseous state is called **sublimation**.

V. *Heating Iodine*

1. Obtain a sample of iodine in a sealed tube. Dip the tube of iodine in a beaker of warm water and observe what happens.

2. Remove the tube and put it immediately into a separate beaker of cold water. Once again observe what happens inside the tube.

(a) What did the iodine in the sealed tube look like?
(b) What did the iodine look like after it was heated in the warm water? After it was cooled in the cold water?
(c) Has a change occurred? Explain your answer.

Follow-up

1. In some of the processes you have just observed something new was produced. In each of these cases, the new material did not resemble the original materials. In fact, if you left the new materials overnight and returned the next day, you would see no change. The reactions that produced them are ''one-way'' changes. Which processes followed this pattern?

2. In other changes, the change is easily reversed by simply reversing the conditions. Name one example of this type of change which you observed in this investigation.

3. When wood is burned, water vapour (or steam), other gases, and charcoal are produced. After the burning, however, you cannot put these materials together again to make wood. Burning wood is an example of a **chemical change**. A chemical change produces a new substance with different physical and chemical properties from the original substance. Which changes did you observe in this investigation that are chemical changes?

4. In a **physical change**, the chemical composition of the original substance remains the same. For example, ice, water, and steam are all the same chemical substance. So a change such as boiling water is a physical change. Often a physical change can be easily reversed. Which changes did you observe in the investigation that are physical changes?

Fig. 5 *When exposed to air for a long time, particularly in an industrial area, the surface of silverware darkens. This process is called* **tarnishing**. *Contact with certain foods also causes silver to tarnish.*

5. Decide whether each of the following is a chemical change or a physical change. Write in your notebook the letters "a" to "p". Beside each letter, write "Ch" (for chemical change) or "Ph" (for physical change).
   (a) iron rusting
   (b) dynamite exploding
   (c) a cigarette burning
   (d) water boiling
   (e) meat barbecuing
   (f) stone being crushed
   (g) sulfur melting
   (h) silver tarnishing
   (i) an egg frying
   (j) sugar dissolving in water
   (k) a cake baking
   (l) leaves changing colour in the fall
   (m) food freezing
   (n) sulfur burning
   (o) paint fading
   (p) glass breaking

6. You may observe some or all of the following during a chemical change:
   • the formation of gas bubbles
   • the formation or disappearance of a solid
   • a permanent change in colour
   • heat or light given off or taken in
   • the production of electricity (Figure 6)

Are these observations sufficient to prove that a chemical change is taking place? (Have you observed any of them in physical changes you are familiar with?)

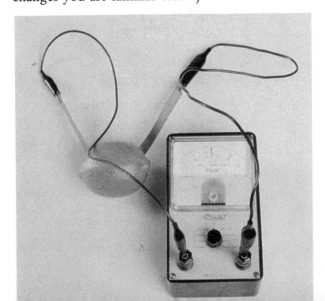

Fig. 6 *Stick a piece of zinc and a piece of copper into the flesh of a lemon. The voltmeter will show you that electricity is produced. Hook up several "lemon cells" in series. See who can get the most electricity!*

| Investigation | 2.3 Element or Compound? |

In investigation 2.2, you worked with quite a few chemicals. In sections 1.8 and 1.9 and investigations 1.10 and 1.11, you learned a lot about classifying substances on the basis of their properties. We will now take another look at ways of classifying chemicals.

First, recall some terms you have learned in earlier courses. A chemical that is not mixed with any other is known as a **pure substance**. In investigation 2.2, the iron filings, sulfur powder, iodine, magnesium ribbon, and marble chips were all examples of pure substances. If you put pure substances together, the result is a **mixture**. Recall in part I the mixture of iron and sulfur before it was heated. You were able to separate the iron from the sulfur.

A special kind of mixture is a **solution**. It looks like a pure substance, but it is not. There are many kinds of solutions, but the commonest is made by dissolving a solid in a liquid. (The most common liquid is water.) Recall the calcium chloride and sodium hydroxide solutions in part III.

A common way of classifying pure substances is to group them into **elements** and **compounds**. A compound is a pure substance that can be broken down into two or more simpler substances. Water is a good example. We can break it down to oxygen and hydrogen by passing electricity though it. A pure substance that cannot be broken down into simpler pure substances is an element. Oxygen and hydrogen are both elements. In investigation 2.2, you used elements, such as iron, sulfur, and iodine, and compounds, such as marble. But how can we tell whether a pure substance is an element or a compound?

Materials

safety goggles
Bunsen burner
ring stand
ring clamp
pipe stem triangle
crucible
tongs
balance
copper powder or copper turnings

Method

1. Examine a sample of pure copper and note its properties.

2. Put 3 g to 4 g of copper into a crucible. Weigh the crucible and contents.

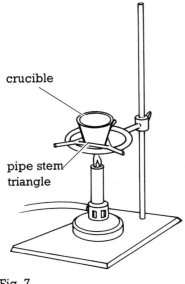

crucible

pipe stem
triangle

Fig. 7

3. Heat the crucible strongly (Figure 7). Examine the contents from time to time, watching for changes in their appearance. Continue heating until no further changes seem to be taking place.

4. Allow the crucible to cool, then reweigh it.

5. Remove a small quantity of the contents from the crucible. Examine the material and note its properties. (Keep the contents of the crucible for the next investigation.)

Follow-up

1. How did the copper change in appearance as a result of heating?

2. Was the change that occurred physical or chemical?

3. Why does the substance in the crucible weigh more after the heating than before?

4. Consider the "new" substance in the crucible. Is it an element or a compound? Explain your answer.

5. Substances that combine in a chemical reaction are called **reactants** (or **reagents**). Substances that are formed as a result of a reaction are **products**. In question 4, you decided whether the product of one reaction was an element or a compound.
   (a) What are the reactants in this reaction?
   (b) Are the reactants elements or compounds? How would you determine if your answer is correct?

A reaction in which two substances react to form just one substance is a **combination** or **synthesis** reaction. The word "synthesis" comes from the Greek word *syntithenai*, meaning to put together.

6. We can describe **combination** and other reactions in a short-hand way by means of a **word equation**. In a word equation, an arrow separates the reactants from the products. The names of the reactants appear on the left, separated by a + sign, and the names of the products appear on the right, separated by a + sign. (If there is only one reactant or product, there is no need for a + sign on that side of the arrow.) For the reaction of oxygen and hydrogen to produce water, the word equation is as follows:

$$\text{hydrogen } + \text{ oxygen } \rightarrow \text{ water}$$

If the product of the reaction in this investigation is copper oxide, what is the word equation for the reaction you observed?

7. What rule of thumb will help you determine whether a reaction results in the formation of a compound?

33

8. Based on this rule of thumb, how would you determine whether the following changes resulted in the formation of a compound?
   (a) Magnesium burns in air.
   (b) Silver tarnishes.
   (c) Iodine vaporizes.
   (d) Iodine vapour becomes a solid.
   (e) Iron reacts with sulfur.

CHEMICAL TIDBITS

CUTTING DOWN ON THE HOUSEWORK!

Many homes have windows with aluminum frames. The surface of the aluminum looks dull (Figure 8). This is because the following chemical change takes place when aluminum is shiny:

aluminum + oxygen → aluminum oxide

The aluminum oxide coating prevents further corrosion, or destruction, of the aluminum surface beneath. So do not waste your time trying to shine aluminum frames.

Aluminum oxide is very different from the destructive iron oxide, **rust**. Rust is called a destructive oxide because it falls off the surface of iron and does not protect the metal underneath. Rust is the product of this reaction:

iron + oxygen → iron oxide

Fig. 8 *It is all right for aluminum to look dull!*

| Investigation | ## 2.4 Taking Apart a Compound |

To **analyze** means to take something apart. The word comes from the Greek word *analyein,* meaning to break up.

If an element ''gains weight'' when it combines with another element to form a compound, what do you suppose happens when we take a compound apart? Let's find out by taking apart copper oxide, the compound that was formed in investigation 2.3.

Materials

safety goggles
Bunsen burner
2 test tubes
a glass tube bent and inserted through a one-hole stopper
mortar and pestle
balance
product from investigation 2.3
charcoal powder
limewater

Method

1. Take the product (copper oxide) from investigation 2.3 and grind it gently with a mortar and pestle.

2. Add about 2 g of powdered charcoal to the copper oxide. Mix the substances and transfer the mixture to a test tube. Weigh the test tube and its contents.

3. Pour a few millilitres of limewater into the second test tube. Connect the first test tube via a stopper and glass tube to this second test tube (Figure 9).

4. Heat the mixture in the first test tube and observe what happens to the limewater in the second test tube.

5. When the bubbles stop forming, turn off the heat and immediately remove the test tube of limewater. (If you do not, the liquid may be sucked into the hot test tube, which may crack as a result.)

6. Allow the product to cool, then weigh the first test tube and contents. Examine the contents.

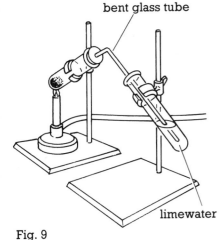

bent glass tube

limewater

Fig. 9

Follow-up

1. (a) What happened to the copper oxide when you heated it with charcoal?
   (b) Is this change chemical or physical?

2. (a) What happened to the limewater when gas bubbled through it?
   (b) Was this change chemical or physical?
   (c) Identify the gas that bubbled through the limewater.

3. (a) What was left in the first test tube after the reaction?
   (b) How do you know?

4. (a) How does the mass of the product in the first test tube compare to the original mass of the copper oxide?
   (b) How can you explain this difference?

5. Charcoal is a type of carbon. In question 2(c), you identified the gas that bubbled through the limewater. Carbon is part of this gas.

   (a) What element did the carbon combine with to form this gas?
   (b) Where did this element come from in the reaction?

6. (a) Write a word equation for the reaction between copper oxide and charcoal.
   (b) Which reactants and products are elements and which are compounds? Explain your answer.

7. In investigation 2.3, you met the term "synthesis reaction." The opposite of a synthesis is a **decomposition**, in which one substance breaks down into two or more substances. An example is the decomposition of water into hydrogen and oxygen:

$$\text{water} \rightarrow \text{hydrogen} + \text{oxygen}$$

   Another common type of reaction is a **single displacement** (sometimes called "single replacement" or just "replacement"). This type of reaction is of the general form:

$$\text{element}_1 + \text{compound}_1 \rightarrow \text{element}_2 + \text{compound}_2$$

   The following equation is an example:

$$\text{zinc} + \text{silver chloride} \rightarrow \text{silver} + \text{zinc chloride}$$

   Notice that zinc replaces silver.

   (a) Examine the word equation in question 6(a). Is this reaction a single displacement or a decomposition reaction?
   (b) Look at your answers to question 6(b). If you were unsure about any of your answers to that question, can you now make a better decision on the basis of the reaction type?

8. What is a compound?

9. What is an element?

10. Classify each of the following reactions as a synthesis, decomposition, or replacement.
    (a) iron + sulfur → iron sulfide
    (b) silver oxide + hydrogen → silver + hydrogen oxide (water)
    (c) magnesium + oxygen → magnesium oxide
    (d) aluminum oxide → aluminum + oxygen
    (e) iron + oxygen → iron oxide

11. Write a combined report on investigations 2.3 and 2.4.

CHEMICAL TIDBITS

THE STEEL INDUSTRY

Fig. 10 *Making steel.*

Note that the names of the oxides are not chemically correct.

Iron ores used in iron and steel production are mainly iron oxides. Concentrated iron ore pellets, some coke (a source of carbon), and some limestone are loaded into giant ovens. These substances are heated from below with hot air. As they melt, they mix together.

The overall reaction between iron oxide and carbon can be summarized as follows:

iron oxide + carbon → iron + carbon oxide

The limestone in the mixture breaks down to form lime, which combines with unwanted parts of the ore to form a by-product, called slag. Molten slag and molten iron collect at the bottom of the oven. The slag floats on the iron, so the two liquids can be drawn off separately. The iron is valuable, and the slag is not wasted. It is used in paving roads and making concrete blocks, for example.

| Investigation | *2.5 Controlling Chemical Changes* |

In investigations 2.3 and 2.4, you used heat to start chemical reactions. There was no visible chemical change in each case until the heating began. If any reaction was going on before you began heating the substance, the reaction was so slow that you could not detect it. Chemical changes that require heat are very common. Think of cooking, for example. Also, to ripen fruit, we keep it in a warm spot. The heat speeds up the chemical changes in the fruit.

If heating helps chemical reactions to happen more quickly, it seems logical that cooling should then help slow them down. This is, in fact, true. We cool food in refrigerators to slow down chemical changes so that the food keeps longer. At even lower temperatures, for example, in a freezer, we can almost stop the changes. But heating and cooling are not the only ways of speeding up and slowing down chemical reactions.

Consider the common drug aspirin, which you have probably taken to relieve a headache or fever. The reaction that produces aspirin is usually very slow and can be speeded up by using concentrated sulfuric acid as a **catalyst**. A catalyst is a chemical that speeds up a chemical reaction but does not get used up itself. We can also add chemicals to slow down a chemical reaction such as the spoilage of food. Such chemicals are called **inhibitors**. They are used to lengthen the shelf life of many foods and other products. We will now examine the effect of a catalyst on a chemical reaction.

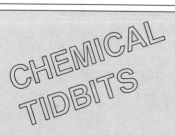

THE ASPIRIN STORY

As early as 1763, it was reported that a brew made by boiling willow bark in water helped people who were in pain (Figure 11). The active ingredient in willow bark was later found to be salicylic acid. In some people salicylic acid had bad side effects, such as stomach pain and bleeding. In 1899, Felix Hoffman of the Bayer Company in Germany made a related chemical, acetyl salicylic acid (ASA). It was very successful in relieving pain and had fewer side effects. The company called it ''aspirin.''

Aspirin is a most useful product. When you have a cold, the most common prescription is to go to bed, drink lots of fluids, and take an aspirin. Aspirin reduces fever, pain, and inflammation. It is used daily by people with arthritis and gout. The drug helps stroke

The name ''aspirin'' comes from ''a'' in acetyl and ''spi-rin,'' which was the common name for the ingredient in willow bark.

Materials

safety goggles
wooden splints
Bunsen burner
spatula
test tube
matches
test tube rack
hydrogen peroxide solution (30 volume)
manganese dioxide

Hydrogen peroxide solution
damages the skin.

Method

1. Place a 1-cm depth of hydrogen peroxide solution in a clean test tube. Do you observe any change taking place in the solution?

2. Using a dry spatula, measure out about a pea-sized quantity of manganese dioxide powder. Add this to the solution in the test tube. What do you observe?

3. When the reaction is occurring fairly vigorously, bring a glowing splint up to, and just inside, the mouth of the test tube. Do not place the splint too far into the tube. What happens when the gas produced comes in contact with the glowing splint?

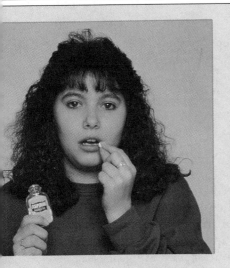

victims by reducing the clotting ability of blood and preventing strokes. However, this reduction in blood clotting could cause stomach bleeding in people who suffer from certain stomach ulcers. People who take too much aspirin experience buzzing or ringing in their ears and temporary hearing loss. An overdose can cause the death of a child or even an adult. Cats are poisoned by small amounts of aspirin and related chemicals.

Aspirin should never be stored in a humid place or for too long. It decomposes after a while. One of the products is salicylic acid, which is not very dangerous but does cause stomach irritations.

Fig. 11 *Aspirin is a very common pain reliever.*

Follow-up

1. What happened to the hydrogen peroxide solution when you added the manganese dioxide?

2. Was the manganese dioxide used up in the reaction?

3. What role did the manganese dioxide play in the chemical change you observed?

4. (a) What happened when you put the glowing splint at the mouth of the test tube?
   (b) What gas was present?

5. The word equation for the reaction that occurred is as follows:

   hydrogen peroxide → water + oxygen

   (a) What type of chemical reaction is this?
   (b) Decide whether each reactant and product is an element or compound.

Since manganese dioxide is not used up in the reaction, it is not shown in the word equation.

6. (a) What is a catalytic convertor?
   (b) What kind of reaction does it speed up?
   (c) Why is it beneficial to the environment to have catalytic convertors installed in automobiles?

7. When liquid vegetable oils are converted to margarine, the metal nickel is used in the conversion process of **hydrogenation**. Do you think nickel serves as a catalyst or an inhibitor?

We **hydrogenate** a substance when we react it with hydrogen.

8. (a) What sort of foods contain inhibitors to prevent spoilage?
   (b) What can happen to you if you eat food that has begun to spoil?

9. You want to store some potatoes for the winter and you have a choice of storage spaces — a warm kitchen or a cool basement. Which room do you choose and why?

| *Investigation* | *2.6 Controlling Physical Changes* |

In investigation 2.5, you considered ways of controlling chemical changes. But what about physical changes? Are there ways of preventing them or bringing them about? Think of your home and your day-to-day activities. Draw up a list of the various means you have encountered for bringing about or preventing physical changes. Discuss your list with your group, then investigate.

Materials

safety goggles
two 250-mL beakers
2 thermometers
stirring rod
water
crushed ice
table salt

Method

1. Mix about 50 mL of cold water with about 100 mL of crushed ice in a beaker.

2. Repeat step 1 using a second beaker, then stir about 20 g of table salt into the mixture.

3. Record the lowest temperature of the contents in each beaker.

Follow-up

1. The lowest temperature you recorded for each ice-water mixture is the **melting point** of the ice (and **freezing point** of the water).
   (a) What is the effect of the added salt on the melting point of ice?
   (b) What practical application does this effect have?

Graphite is a form of carbon.

2. **Lubricants,** such as grease and graphite, are materials that help the moving parts of machines operate more smoothly.
   (a) What physical changes are lubricants intended to prevent?
   (b) What must they reduce in order to be successful?

3. Section 1.12 and investigation 2.4 featured the Chemical Tidbits "The Making of Glass" and "The Steel Industry." Identify physical changes involved in these production processes.

4. Modern buildings are insulated to reduce heat flow. Find out what kinds of insulating materials are used and why they are effective.

5. Pour two samples of warm water, at the same temperature, into a glass beaker and a styrofoam cup.
   (a) Which container provides better insulation?
   (b) How do you know?

6. Stainless steel contains a small amount of chromium.
   (a) What advantages does stainless steel have over other varieties of steel?

(b) Separate these advantages into those that depend on physical characteristics and those that depend on chemical characteristics.

7. Pyrex® glass is used not only in laboratory glassware but also in making see-through cookware. What advantage does Pyrex® glass have over other varieties of glass?

8. Nowadays much of the clothing we wear and the linen we use is made of a blend of cotton and synthetic fibres. Why do the manufacturers of these fabrics add synthetic fibres? Do these additives improve or reduce the quality of the fabrics?

CHEMICAL TIDBITS

Pure metals taken from ores are generally mixed with other metals to form **alloys**, which have improved physical properties. For example, aluminum is added to steel to make a lighter alloy more useful for building aircraft. Steel and stainless steel are themselves alloys.

Fig. 12 *A type of transformer.*

AN EXPENSIVE PHYSICAL CHANGE

Undesirable chemical changes, like the rusting of iron, rotting of wood, and spoiling of food, cost our society a great deal of money. So do undesirable physical changes such as "metal fatigue" in aircraft. A less well-known example is associated with the transmission of electricity. As you will learn in later chapters, efficient transmission depends on the use of "transformers" (Figure 12). **Cavitation**, which is the wearing away of the surface of the steel used in transformers, is a serious problem. It is caused by the flow of oil around the steel. Samples of the oil are taken periodically and tested in laboratories. Sensitive meters are able to detect tiny flecks of iron suspended in the oil.

Cavitation reduces the efficiency of electrical transmission. Steel loss costs amount to $1 500 000 a year in North America. Hydro-Québec, however, has developed a repair alloy called Hydroloy HQ913 that extends the life of the metal four to ten times longer. This alloy contains iron, cobalt, manganese, carbon, nitrogen, and silicon in specific quantities.

STAINLESS STEEL COOKWARE

Closer to home, there are other examples of technological improvements that prevent undesirable physical changes or bring about desirable ones. For example, pots and pans made of stainless steel are very sturdy and easy to clean (Figure 13), but they have a tendency to heat unevenly, causing food to cook unevenly and burn in spots. Copper and aluminum tarnish rapidly and are much softer metals, but they are excellent conductors of heat. To preserve the good properties of stainless steel and improve on its ability to conduct heat, a layer of copper or aluminum is sandwiched between layers of steel.

Brainstorm with a group of your classmates and come up with a list of ways in which technology has improved the physical properties of materials we use daily. Share your list with the class.

Fig. 13 *Stainless steel cookware.*

ONE TECHNOLOGY LEADS TO ANOTHER!

The action of soap in most household water leaves a scum on washed clothes. If the clothes are sun dried, this scum traps a layer of water, so there are no static charges on the fabrics. However, if the washed clothes are dried in a clothes dryer, the fabrics feel harsh and contain static charges, especially in winter. That's why fabric softeners were designed! They may be added during or after the wash cycle, or even in the dryer, depending on the chemicals in the product.

Fabric softeners leave a greasy layer on the fabrics, making them feel soft. This greasy layer also traps a layer of water, reducing the static charge. However, if you use fabric softener with every wash, the greasy layer builds up and attracts grime, and the water-absorbing ability of the fabrics is lowered. It is a good idea to do a wash without fabric softener every now and again.

Hair conditioners are like fabric softeners in many ways. Describe how you think they work. What disadvantages do hair conditioners have?

Charlie

"..And for those doing their laundry tomorrow, the Weather Bureau has issued a severe 'static cling' advisory..."

2.7 The Impact of Change

We began this chapter with some thoughts about the change of season. What kind of chemical and physical changes are involved in each of the following?
- the sprouting of a tulip bulb
- the melting of ice on rooftops
- the growing and cutting of grass
- the use of salt on our sidewalks and roads in winter
- the preparation of a compost heap to provide fertilizer for next year's garden
- the leaves changing colour and falling
- the creation of potholes in our streets in the early days of spring

Can you think of other ways in which physical and chemical changes have an effect on our daily lives?

Some changes may have a bad effect on our environment. Did you know, for example, that Canada leads the industrialized world in energy consumption? That means we burn more fuel and produce more carbon dioxide per person than the United States, Australia, New Zealand, the Netherlands, or any other industrialized country does! The growing amount of carbon dioxide in the atmosphere has been linked to a trend in global warming. Many scientists believe that this will have a negative effect on the world's environment. How would an increase in the global temperature affect the polar icecap? How, in turn, would global warming affect our coastlines? How would it affect agriculture in Canada and in other parts of the world?

The Changing Atmosphere Conference held in Toronto in 1988 targeted a 20% reduction in carbon dioxide emissions by the year 2005. Some politicians believe such a reduction would bring about a serious disruption in the Canadian economy.

If one of the measures taken to reduce the carbon dioxide in the atmosphere were to limit transportation by cars and trucks, how do you think this would affect the economy? If another measure were to discourage the destruction of forest lands, which play a large role in converting carbon dioxide to oxygen, how might this affect our economy?

These are serious questions. In small groups discuss both sides of the issue. Can you imagine a solution that would benefit both sides? Discuss your group's suggestions with the class.

Fig. 14 *Rain forests are factories for the conversion of carbon dioxide to oxygen. The loss of rain forests to agricultural or industrial purposes has been blamed for part of the increase in global warming.*

P O I N T S · T O · R E C A L L

- Fuel, oxygen, and a sufficiently high temperature are needed for burning to occur.
- In a physical change, the chemical composition of the original substances remains unchanged, and no new substances are formed.
- In a chemical change, new substances are formed.
- A word equation uses words to describe the chemicals used (reactants) and the chemicals produced (products) in a chemical reaction.
- In a synthesis reaction, elements or smaller compounds combine to form a single product.
- In a decomposition reaction, a single reactant breaks down into elements or smaller compounds.
- A single replacement reaction has the form:
  $element_1$ + $compound_1$ → $element_2$ + $compound_2$
- A catalyst is a substance that speeds up a chemical reaction and is itself not used up in the reaction.
- An inhibitor is a substance that slows down a chemical reaction and is itself not used up in the reaction.

R E V I E W · Q U E S T I O N S

1. What is a chemical change?
2. What is a physical change?
3. Write down all the steps you would take if you woke up and smelled smoke in your bedroom. Give a reason for each step.
4. List three safety precautions you would take in your home to prevent a fire starting. Give a reason for each precaution.
5. What is spontaneous combustion? Give an example of a situation that could result in spontaneous combustion.
6. What is the main difference between a chemical change and a physical change?
7. List at least four observations that may tell you that a chemical change is occurring.
8. Decide whether each of the following is a chemical change or a physical change.
   - (a) a candle burning
   - (b) ice cream freezing
   - (c) food rotting
   - (d) ice melting
   - (e) jelly dissolving
   - (f) bread toasting
9. What is a word equation?

10. (a) Why is it not safe to look directly at the flame of burning magnesium?
    (b) Why should you avoid exposing your eyes to non-electronic flashbulbs for too long?
11. What are the three types of chemical reactions described in this chapter? Give an example of each type.
12. Why is it a waste of time to polish aluminum window frames to make them shiny?
13. What is a catalyst? Why is a catalyst never included in the word equation for a reaction?
14. What is an inhibitor? Give an example of the use of an inhibitor in everyday life.
15. Complete the following word equations. (Please do not write in this textbook.)
    - (a) aluminum + oxygen →
    - (b) iron oxide + aluminum →
    - (c) zinc + chlorine →
    - (d) water (hydrogen oxide) →
    - (e)   +   → sulfur dioxide
    - (f) silver + sulfur →
    - (g) sodium oxide →
16. Classify each reaction in question 15 as a synthesis, decomposition, or single replacement.

3

THE ATOMIC MODEL

The colours of fireworks related to the energies of electrons in atoms.

CONTENTS

3.1 What Is Matter?

Have you ever wondered what the stars are made of? what the earth, the sea, or you yourself are made of? Think of all the many different kinds of matter with their vast array of different properties. Consider also that all matter can, under the right circumstances, be changed from one form to another.

For example, water can be frozen to make ice, or boiled to make steam. Each of these physical changes is easily reversed because ice and steam are still forms of water. What is it, though, about each form that makes it water? It is more than appearance, surely. In some ways, the three forms look similar (they are all colourless, for example), but in other ways they look different.

In contrast to the freezing and boiling of water, some changes cannot be reversed. When a piece of wood burns, the matter present in the wood changes permanently into different types of matter. The gases given off and the ash that remain come from the matter in the wood.

But why is it that matter can undergo different types of changes, some of which just change the matter's form while others produce new types of matter?

The ancient Greeks asked the same question and their best thinkers and teachers came up with some ideas on how to describe the nature of matter. They called this branch of thinking natural philosophy.

Some natural philosophers imagined that matter was **continuous**, meaning that there are no gaps or breaks in it. We think of a ribbon as continuous since, if it is cut into smaller pieces, the pieces are still ribbon. These philosophers pointed to the way water retains its nature and properties even if it is divided into droplets.

Other natural philosophers imagined that matter is **discontinuous**, meaning that it contains pieces with gaps between them. To them, matter was composed of tiny units or building blocks, as sand is composed of grains, or salt is composed of crystals. For over a thousand years, there were supporters of both views of matter.

These views of matter are examples of models, which are concrete images that help people to think. When you want to think about something that no one can see, then you imagine what it might look like in order to explain the way it acts. This imaginary picture is called a model. For instance, what is your model of a Sasquatch? What is your model of the structure of matter? Do you think matter is continuous or discontinuous? What are the reasons for your choice?

Around 450 B.C., Democritus was among the first to question the continuity of matter. He asked students to imagine a block of silver and then to imagine dividing it into two equal pieces. Each half still has the properties of silver. If one of those blocks were split in half, the smaller blocks would continue to have all the properties of silver. If one of these smaller blocks were halved again and again, you would eventually arrive at a very small particle that would still have all the properties of silver. Democritus coined the term "atomos," meaning indivisible or uncut, to describe these extremely small particles. The idea that matter is made up of very small particles that are not divisible is known as atomism.

Aristotle (384–322 B.C.), the tutor and friend of Alexander the Great, rejected atomism. Aristotle said that, because he could not detect atoms with his senses, he was not willing to accept their existence. Instead, he preferred the idea that matter was made of four elements—fire, air, water, and earth—that the senses could detect. All matter, he claimed, was composed of varying amounts of these four elements. The proportions in which they were combined to produce a substance determined the properties of that substance. Liquids, for example, must have large amounts of water, and fuels must contain large amounts of fire.

*A **model** is a mental picture of something we cannot actually see.*

Questions

1. Do you think that Aristotle's reason for rejecting Democritus's idea of atomism is valid?

2. Why do you think the use of models is so valuable in science?

3. (a) How would you explain the flow of water around a rock using Democritus's model? Using Aristotle's model?
   (b) How would you explain the burning of a log?
   (c) Which model is more convincing to you?
   (d) Which model is easier to use?

3.2 A Model For Chemical Change

In investigation 2.3, you formed black copper oxide by heating copper metal in a crucible in the presence of oxygen gas. What happened to the mass of the material in the crucible during this change? What did that change in mass tell you about what happened to the copper? In investigation 2.4, you reacted the black copper oxide with carbon. What happened to the copper oxide?

Given a supply of wire and coloured styrofoam spheres, decide how to represent each of the following. Draw a picture of each representation in your notebook.

- the smallest particle of the element copper
- the smallest particle of the element oxygen
- the smallest particle of the element carbon
- the reaction between copper and oxygen
- the reaction between copper oxide and carbon

How does your model of the reaction between copper and oxygen explain your observation about the change in mass during the reaction in investigation 2.3? How does your model of the reaction between copper oxide and carbon explain the observation about the change in mass during the reaction in investigation 2.4? Does it explain the change in appearance of the material or how the gas was formed?

Questions

1. Rate your model on its ability to represent the chemical reactions you observed.

2. How could you improve your model so that it does a better job of representing the reaction?

3. What makes for a good model in chemistry? List the qualities that a good model must have.

4. Use coloured styrofoam spheres and wire to represent
   (a) each of these synthesis reactions:
       (i) magnesium + oxygen → magnesium oxide
       (ii) aluminum + oxygen → aluminum oxide

(b) each of these decomposition reactions:
  (i) water $\rightarrow$ hydrogen + oxygen
  (ii) sodium chloride $\rightarrow$ sodium + chlorine

(c) each of these single replacement reactions
  (i) sodium iodide + chlorine $\rightarrow$

  sodium chloride + iodine
  (ii) magnesium + aluminum chloride $\rightarrow$

  magnesium chloride + aluminum

3.3 Dalton's Atomic Model

Although the idea that matter was composed of atoms had been proposed by the ancient Greeks, it was John Dalton, in about 1800, who converted the idea of atomism into a useful, working hypothesis (Figure 1).

Fig. 1 *John Dalton (1766–1844) was the first person to develop a model of the atom, but his idea that the atom was a solid sphere has now been disproved.*

This English schoolmaster became aware of research being done in Europe on the composition of matter. He read a great deal about the work of Antoine Laurent Lavoisier (1743–1794), and others, who showed that pure substances combine in fixed ratios by mass. For example, they showed that it did not matter how much water you formed or how you formed it, water always contains eight grams of oxygen for every gram of hydrogen. It was also shown that other elements combined with oxygen in different, but nonetheless fixed, ratios by mass. Dalton asked himself why this was always the case.

He found a reasonable explanation in Greek philosophy. In 1805, he published "A New System of Chemical Philosophy," in which he suggested the following:

(a) All matter is composed of tiny, indestructible particles called atoms.

(b) Atoms of the same element are identical and have a characteristic mass, but they are different from the atoms of other elements.

(c) Atoms of different elements combine in simple whole numbers to form compounds.

Dalton was a man of regular habits. He took measurements of the daily rainfall and temperature — rather a boring exercise, since the climate where he lived in Manchester, England, changed very little. Each and every week he went lawn bowling. Dalton pictured the atom as a hard sphere, not unlike the bowling balls he rolled down the greens once a week. He never saw an atom — to this day no one has seen one clearly — but he found a useful way of picturing atoms, a model of the atom, that enabled him and others to imagine the structure of matter and to explain its behaviour.

He assigned symbols to the elements and combined symbols to represent compounds. He also prepared a list of atomic masses, or atomic weights as he called them, for 20 elements. He chose hydrogen, the lightest element, as a standard for atomic mass. He assigned a mass of one unit to the hydrogen atom and based the masses of other atoms on the mass of the hydrogen atom (Figure 2). Some of his atomic masses were not very accurate.

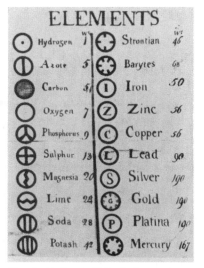

Fig. 2 *Dalton's list of elements.*

Questions

1. Compare Dalton's model to Democritus's model in terms of their similarities and differences.

2. Does Dalton's model meet your criteria for a good model?

3. Use Dalton's atomic symbols to represent each of the following compounds.
   (a) carbon monoxide, containing one atom of carbon for every one atom of oxygen
   (b) hydrogen sulfide, containing two atoms of hydrogen for every one atom of sulfur
   (c) nitrogen monoxide, containing one atom of nitrogen (which Dalton called "azote") for every one atom of oxygen

(d) ammonia, containing three atoms of hydrogen for every one atom of nitrogen

(e) sulfur dioxide, containing two atoms of oxygen for every one atom of sulfur

(f) sulfur trioxide, containing three atoms of oxygen for every one atom of sulfur

(g) phosphorus trichloride, containing three atoms of chlorine for every one atom of phosphorus

| *Investigation* | *3.4) The Shocking Nature of Matter* |

Have you ever touched a metal doorknob after walking across a carpet in nylon socks on a cold, dry day? Then you know it can be a "shocking" experience! When you comb or brush your hair and it gets that flyaway look, are you thinking about chemistry? You should be!

Discuss in small groups the phenomena described in the above paragraph. Propose an explanation and share your ideas with the class. How does your hypothesis fit in with the atomic models we have already discussed?

Materials

ring stand
ring clamp
woollen cloth
cotton cloth
2 strips of vinylite plastic
2 strips of cellulose acetate plastic
cotton thread
comb
balloon
flour

Method

1. Rub each of the plastic strips with the woollen cloth.

2. Bring two like strips (either vinylite or cellulose acetate) together and observe what happens.

3. Bring two different kinds of plastic strips together and observe what happens.

4. Attempt to pick up tiny bits of paper with both types of plastic strips. As well, try to pick up other tiny objects, such as pieces of plastic, specks of dust, and grains of flour.

5. Tie one end of the cotton thread to the ring clamp, so that most of the thread is suspended. Observe what happens to the hanging thread when you place each type of plastic strip near it.

6. Repeat steps 2 to 5 using the cotton cloth in place of the woollen cloth. Do you notice anything different?

Follow-up

1. Write down all your observations in a table.

2. Compare your observations with those of your classmates.

3. When you tried to pick up the various tiny objects with the plastic strips, which objects behaved similarly? Which behaved differently?

4. What changes, if any, would you make to your original hypothesis to explain your observations?

5. Predict what would happen if you brought a charged object such as a comb, which you have just used to comb your hair, toward small pieces of paper or plastic. Test your prediction experimentally.

6. You can also charge a balloon by rubbing it on your hair. What would you expect to happen when you bring this balloon into contact with small pieces of paper or plastic strips? What would happen if you brought two charged balloons together? Check your hypotheses by appropriate experimentation.

3.5 Understanding Static

The phenomena you examined in investigation 3.4 were observed by the ancient Greeks as early as 3000 B.C. It was Thales who first recorded that amber, when rubbed with fur, would attract tiny objects.

Amber became prized for its "magical" properties. The Greek word for amber is *elektron*. Any object with such powers of attraction was said to be charged with electricity.

In the seventeenth century, it was discovered that this electricity could pass from one object to another merely by touch. The object touched would then have some of the attracting abilities of the charged object. However, the second object would repel the object from which it acquired the charge.

Evidence showed there were two kinds of charge — one derived from rubbing amber, the other from rubbing glass. These two kinds of charge attracted one another but repelled anything bearing the same charge. It was widely believed that two kinds of electrical ''fluid'' produced these mysterious phenomena. These were curious fluids, however. A charge, it was found, could even be induced in a second object when no direct contact had been made. How could the fluid get into the second object?

Meanwhile, the public was getting a ''charge'' out of all the talk about electricity. In the eighteenth century, people lined up to get a dose of electrical charge dispensed by generators. Inside a generator, chunks of amber were mounted on a wheel that rubbed the amber against brushes made of cat's fur. The charge was considered a remedy for low spirits.

Benjamin Franklin (1706–1790), the American inventor and leading public figure, believed that there was only one electrical ''fluid'' and that this was composed of electrical particles (Figure 3). Every substance had its ''normal'' amount of this fluid. The substance would become charged if extra fluid were added (positive charge) or were lost (negative charge).

Fig. 3 *Benjamin Franklin.*

From the late 1700s through to the 1800s, many scientists contributed to what we know about charged objects. You will now research one aspect in the development of our understanding of what we now call "static electricity."

1. Work in *home groups* of four. Each member of the home group will be given a topic to research. The topics are shown in Figures 4 to 7 below.

Fig. 4 *Charles Coulomb discovered that the force of attraction or repulsion depended both on the quantity of charge in the two objects and the distance between them.*

A **force** is a push or a pull on an object.

Fig. 5 *When Luigi Galvani observed that a severed frog's leg twitched when touched by a wire from a generator of static electricity, he proposed that it was due to "animal electricity."*

Fig. 6 *Michael Faraday determined that the quantity of electricity needed to decompose a compound was directly related to the masses of the elements formed. This discovery led to the idea that there could be an "atom of electricity."*

Fig. 7 *Alessandro Volta repeated Galvani's experiment and determined that fluid in the frog's leg carried the electricity to the frog's nervous system. Volta discovered that a steady flow of moving charges could be produced by connecting strips of copper and zinc and dipping the strips into an acidic solution. In other words, he had made the first battery.*

2. Everyone working on the same topic will form an *expert group* to share information (Figure 8). Make notes during your discussion sessions. Write a summary of your information to take to your home group.

3. Each member of the home group will teach the others in the group what they learned in their expert groups.

4. Each home group must prepare and present a brief paper (preferably typed or word-processed), describing the development of the theory of electricity.

Fig. 8

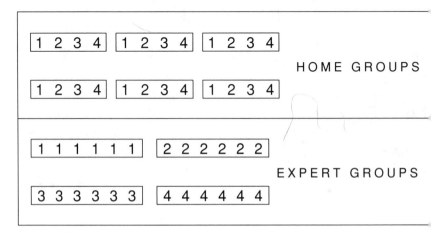

3.6 The Cathode Ray Tube

Evacuated means emptied.

Michael Faraday was not content with exploring the effect of electricity on only solutions and melted substances. He wanted to know what would happen if it were passed through an evacuated glass tube. Unfortunately, his attempts did not meet with success since he lacked a strong enough vacuum pump to remove the air from the tube. By 1875, however, William Crookes was developing a simple tube with an internal pressure of as little as 1.3 kPa.

Pressure is defined as force per unit area.

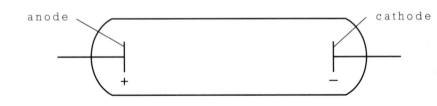

Fig. 9 *A Crookes tube.*

The vacuum obtained in a Crookes tube is about one-hundredth of atmospheric pressure. Thus, the tube is not entirely evacuated.

The SI unit of pressure, the pascal (Pa), is very small. We often use kilopascals (kPa) instead.

Your teacher will demonstrate the properties of the cathode rays using a Crookes tube. It can give you a rude shock if you do not know how to use it safely. Note that the Crookes tube consists of a glass tube with two metal plates called electrodes (Figure 9). The negative plate is called the **cathode**. The positive plate is called the **anode**. If the electricity were switched on with the tube full of air, nothing would change; but if some air is removed using a vacuum pump, a lilac glow appears at first. As more air is removed and the pressure is decreased to around 1.3 kPa, the whole space between the anode and the cathode takes on a pink glow. The colour depends on the gas in the tube. If neon is used instead of air, the glow is orange-red.

In the Crookes tube you will be observing, there is even less gas present. The glass walls of the tube will glow a green colour when the current is turned on. This fluorescent glow in the evacuated tube, beginning at the cathode and going toward the anode, is called a **cathode ray**. No matter what gas the tube is originally filled with, the cathode rays will appear the same green colour. Even the identity of the metal used in the plates has no effect on the cathode rays produced. What does that tell you about cathode rays?

If your cathode ray tube is equipped with a light mica paddle mounted on rails, observe how the paddle behaves when the current is turned on. Toward which plate does the mica paddle move? Does its movement confirm that the rays move from the cathode to the anode? What do you think causes the paddle to turn? What does that tell you about the make-up of cathode rays?

Place a magnet perpendicular to the path of the cathode rays. Toward which pole of the magnet does the cathode ray path bend? What does that tell you about the nature of cathode rays? Applying an electrical field in the same manner would cause the cathode ray to bend toward the positive electrode. Why do you think this happens?

List all the properties of cathode rays that you have observed so far. Are they made of light or of matter? Give reasons for your choice. What is your model of a cathode ray?

J. J. Thomson, an English physicist, (Figure 10) carried out many experiments with cathode rays, including the examination of their behaviour when exposed to electric and magnetic fields. He observed that, in both an electrical field and a magnetic one, the cathode rays were bent or deflected in the direction that a negatively charged particle would also be deflected. As a result of these observations, Thomson proposed that the cathode rays consisted of negatively charged particles. These negatively charged

Fig. 10 *J. J. Thomson (1856–1940).*

particles are called **electrons**, and they are found in all kinds of matter.

Partially evacuated tubes have been adapted for various uses. No doubt you are familiar with neon lights, for example. A neon light is a glass tube containing a little neon gas. This gas glows when electricity passes through it. A television tube is also an evacuated tube; the image on the luminescent screen is created by a beam of electrons. It is possible that if Crookes had not come up with a workable cathode ray tube, we might not have television today. X-ray machines also depend on cathode rays (Figure 11).

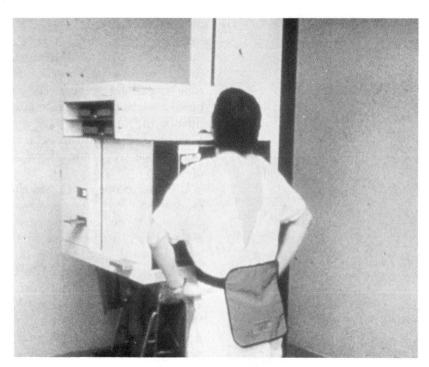

Fig. 11 *X-rays are very important in medicine. They are generated by cathode rays aimed at targets of copper or molybdenum.*

Questions

1. What are the characteristic properties of the particles that make up cathode rays?

2. If electrons come out of matter, then atoms cannot be as indivisible as Dalton thought they were. If electrons are removed from an atom, what can we say about the charge on the remainder of the atom?

3. Write a brief letter to John Dalton describing what you have just observed in the demonstration of the cathode ray tube. Make suggestions to update his atomic theory.

4. Using reference materials as a resource, describe each of the following:
    (a) how sodium vapour lamps operate
    (b) how X-rays are generated

3.7 The Discovery of Radioactivity

Suppose you had an unused roll of film which you placed in a desk drawer alongside an interesting rock that you had collected on a camping trip. One day, you decide that you want to load your camera to take pictures and you grab the film from the desk drawer. To your surprise, the film has been completely exposed, although the container in which it came has not been opened! What could have exposed the film? What kind of light could penetrate the unopened film cartridge? This is the same puzzle that the French scientist Henri Becquerel faced in 1896 (Figure 12).

Becquerel concluded that uranium-containing rock gave out some kind of rays or **radiation**. This could penetrate several thicknesses of black paper and expose the photographic plate that he placed in the same drawer. Uranium was said to be **radioactive**. The rays emitted from this element were first called Becquerel rays.

Fig. 12 *Henri Becquerel, who discovered radioactivity.*

Photographic plates were earlier versions of photographic film. They were made of coated glass.

Fig. 13 *Marie and Pierre Curie, who discovered two radioactive elements, namely, polonium and radium.*

An **ore** is a type of rock we can mine to get a useful material. For example, we get iron from iron ore.

The element polonium was named for Madame Curie's homeland of Poland, and radium for the radiation it emitted.

Shortly after the discovery of radioactivity, Marie Sklodowska Curie (1867–1934) and her husband Pierre devoted several years to studying this new phenomenon, and began a search for other radioactive substances (Figure 13). They proved that the radioactivity of any pure uranium compound, regardless of its physical state, is exactly proportional to the amount of uranium in it.

After several years of hard work, Madame Curie was able to isolate, from a uranium ore called pitchblende, two new and radioactive elements, which she named polonium and radium. For her work, she was awarded two Nobel prizes, one in physics, the other in chemistry. (The physics prize was actually awarded not only to Madame Curie but also to her husband Pierre and to Henri Becquerel.)

Questions

1. How did the discovery of radiation change scientists' ideas about the structure of matter?

2. Using available references, prepare a report on the impact of the Curies on the use of radioactive substances in medicine.

| *Investigation* | ## 3.8 Observing Radiation |

A cloud chamber is a device used to observe radiation. The device consists of a hollow glass cylinder cooled by dry ice (frozen carbon dioxide) and containing water or alcohol vapour. When additional water or alcohol is forced into the cylinder's chamber, the fluid causes the air inside the cylinder to cool rapidly and to become supersaturated with vapour, as the inside of a cloud is supersaturated with water vapour. When radiation travels through the cloud chamber, it leaves trails in the fog behind it (Figure 14).

Supersaturated means "holding more than it should."

Fig. 14 *Particle trails in a cloud chamber.*

Materials

as provided by your teacher

Method

Follow your teacher's instructions on the use of the cloud chamber and radioactive substances.

Using a simplified cloud chamber and some radioactive material, observe the trail patterns in the fog created by the radiation.

Follow-up

1. How many different types of trails did you observe in the cloud chamber?
2. Using reference materials as a source of information, find out how the radiation created those trails.

3.9 Types of Radiation

Fig. 15 *Ernest Rutherford, who developed the nuclear model of the atom.*

Soon after Becquerel's discovery of radioactivity, a New Zealand-born scientist named Ernest Rutherford (1871–1937), carried out experiments that detected three different types of radiation (Figure 15). These types are named alpha (α), beta (β), and gamma (γ) rays. Rutherford was, for a period in his life, a professor of physics at McGill University in Montreal.

Alpha, beta, and gamma are the first three letters in the Greek alphabet. Where do you suppose our word "alphabet" comes from?

These three types of radiation are described in the following table. Notice that alpha and beta radiations consist of streams of particles.

Radiation type	Symbol	Charge	Nature
alpha particle	α	2+	charged helium atom
beta particle	β	1−	high energy electron
gamma ray	γ	0	very high energy light

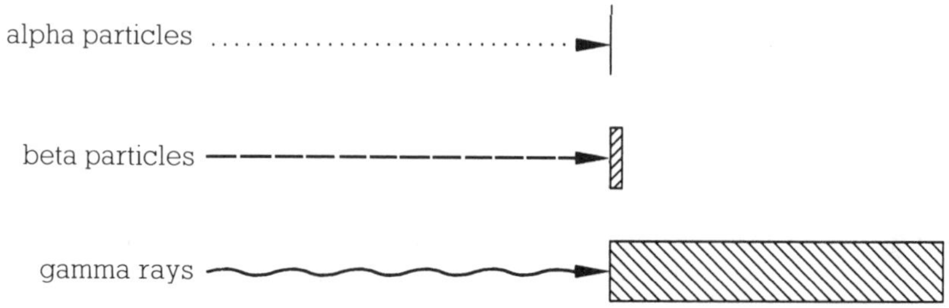

alpha particles .... ▶

beta particles ─ ─ ─ ─ ─ ─ ─ ─ ─ ─ ▶

gamma rays ∿∿∿∿∿∿∿∿∿∿ ▶

Fig. 16 *Gamma rays are much harder to stop than beta or alpha particles.*

The three types of radiation are highly energetic. They can change or destroy the substances they pass through. Cells of living organisms exposed to radiation may be badly damaged or die.

Questions

1. How is it possible for an atom that is supposed to be neutral and indivisible to emit charged radiation? Suggest a possible explanation.

2. Use reference sources to find out which type of radiation caused each type of track observed in the cloud chamber (investigation 3.8).

3.10 Thomson's Model of the Atom

Can the Dalton model of the atom be used to explain the existence of electrons? Can this model explain radioactivity? How can the Dalton model of the atom be modified so that it can explain these phenomena? Suggest an improvement to the Dalton model that would enable you to explain both the existence of electrons in all matter and the radioactivity of certain elements.

J. J. Thomson's studies of cathode rays led him to picture the atom as a positively charged sphere with negative electrons embedded in it, much like raisins in a plum pudding or like seeds in a watermelon (Figure 17). He thought that electrons, like the raisins or the seeds, could be easily removed from the rest of the atom. Why is this easy removal of electrons a necessary feature of the Thomson model? In what way is this model similar to Dalton's model? In what way is this model different from Dalton's model?

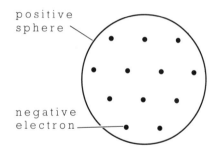

positive sphere

negative electron

Fig. 17 *Thomson's model of the atom.*

Questions

1. How does your revised model of the Dalton atom compare to Thomson's model?

2. Does Thomson's model meet your criteria for a good model?

3. Does Thomson's model explain *all* aspects of radioactivity?

4. What other currently known phenomena cannot be explained by Thomson's model?

5. How would you improve Thomson's model in order to be able to explain these phenomena?

Investigation	## 3.11 A Simulation of the Rutherford Experiment

Materials

flat surface, such as a tabletop
cork board
push pins
a marble
chalk

Method

1. In the centre of a cork board, set up an orderly array of push pins in a square (Figure 18a).

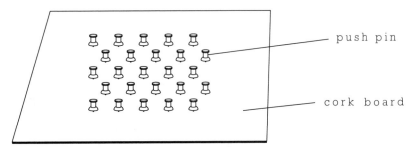

Fig. 18 (a)

2. Gently roll a marble into the array and mark with chalk where the marble emerges from the square.

3. Repeat the previous step 19 more times using approximately the same entry point for the marble and the same initial direction of travel.

4. Group the pins closely together, so that they lie in the path of the marble. Repeat the experiment.

5. Repeat step 4, but first remove all but three of the pins. Leave these three close together.

Follow-up

1. (a) For each of the different arrays you tested, how many times did the marble emerge at a point directly opposite its point of entry to the array?
   (b) Convert your answers to percentages:

$$\frac{\text{number of times directly opposite}}{20} \times 100$$

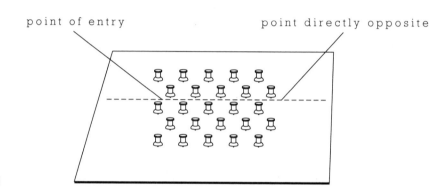

point of entry   point directly opposite

Fig. 18 (b)

2. Imagine a line between the point of entry of the marble and the point directly opposite (Figure 18b). For each array, determine each of the following.
   (a) the number of times the marble emerged less than 90° from the line
   (b) the corresponding percentages

3. Imagine again the same line as in question 2. For each array, determine each of the following.
   (a) the number of times the marble emerged more than 90° from the line
   (b) the corresponding percentages

4. What caused the marbles to change path
   (a) by less than 90°?
   (b) by more than 90°?

5. Did the arrangement of the push pins make a difference in the number of times the marble was deflected? If so, in what way?

3.12 Rutherford's Model of the Atom

In 1911, Ernest Rutherford performed an experiment similar to the one in investigation 3.11. Instead of marbles, he used alpha particles from a sample of radium. He shot these alpha particles at a thin gold foil, and was not surprised that most of the alpha particles passed straight through the gold foil and struck a detector screen on the other side of the foil. He expected this to happen because he had accepted the atomic model of his teacher, J. J. Thomson. Why could he predict, on the basis of the Thomson model, that positively charged alpha particles would not be deflected from their path by the gold foil?

The paths of some of the alpha particles were slightly deflected. Based on your observations in the simulation, what do you think caused the alpha particles to change direction?

Rutherford said of these results, "It was almost as incredible as if you fired a 15-inch shell at a piece of tissue paper and it came back and hit you." (15 in = 37.5 cm)

To Rutherford's great surprise, a very small number of the alpha particles bounced right back from the foil. Would the Thomson model have predicted that? Why not? What do you think caused the alpha particles to bounce back?

How would you revise the Thomson model to take into account the findings of the Rutherford experiment? Draw a diagram of your revised atomic model and show what caused the alpha particles to change course.

Rutherford suggested that the results of his experiments could be explained if the atom had a small, positively charged **nucleus** at its centre. Alpha particles, which are positively charged, would be deflected by repulsion if they happened to pass near a positively charged nucleus (Figure 19). Since most alpha particles were not deflected, the nucleus must be rather small compared to the size of the entire atom. Rutherford in fact calculated that the radius of a gold atom is about 100 000 times larger than the radius of its nucleus. The atom must then be mostly empty space in which the electrons move around the nucleus. What would cause a very small number of alpha particles to bounce back?

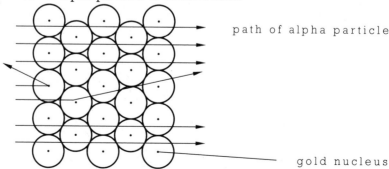

path of alpha particle

gold nucleus

Fig. 19 *Rutherford's experiment showed the scattering of alpha particles by gold nuclei.*

RUTHERFORD IN MONTREAL

CHEMICAL TIDBITS

At age 27, Ernest Rutherford was appointed professor of physics at McGill University (Figure 20). During his nine years at McGill, he worked with a highly gifted English chemist named Frederick Soddy. They published their famous "Spontaneous Transmutation Theory of Radioactivity" in 1903. Their theory stated that, when radioactive elements produce radiation, they are transformed into other chemical elements. The elements formed are generally radioactive themselves and they subsequently break down into other elements. After nine years of hard work at McGill University, and some fifty publications, Rutherford moved to the University of Manchester in England and, in 1908, was awarded the Nobel prize in chemistry. Since he was regarded as the greatest experimental physicist of his day, he was more than a little disappointed to have received this great honour in the field of chemistry!

Fig. 20 *Ernest Rutherford who studied the transformation of radioactive elements into other elements.*

Questions

1. How does the revised Thomson model compare with the Rutherford model?

2. How does the Rutherford model affect our view of matter and its structure?

3. Positive and negative charges attract one another, yet the negative electrons do not fall into the positive nucleus. How could Rutherford's model be adapted to explain this observation?

4. Does Rutherford's model of the atom meet your criteria for a good model?

5. What currently known phenomena are not explained by the Rutherford atom?

6. How could Rutherford's model be adapted to explain them?

3.13 The Bohr-Rutherford Model of the Atom

The smallest atom is that of the element hydrogen. Nearly all the mass of the atom is concentrated in its nucleus. The masses of all other atomic nuclei can be thought of as multiples of the mass of the hydrogen nucleus. The positively charged particle at the centre of the hydrogen atom is called the **proton**. It has what scientists refer to as one unit of positive charge. The quantity of electrical charge on a proton is the same as that on an electron; just the signs are different. How must the number of protons and electrons compare in order to maintain the electrical neutrality of the atom?

Fig. 21 *James Chadwick, who discovered the neutron.*

Fig. 22 *Neils Bohr, who proposed a model of electron arrangement.*

The English physicist, James Chadwick, a former student of Rutherford, received the Nobel prize in 1935 for discovering the existence of a third subatomic particle, the **neutron** (Figure 21). The neutron has no charge, as the name suggests, but it does have a mass similar to that of the proton. (Both the proton and neutron are much heavier than the electron.)

One of the puzzles that the Rutherford model of the atom could not explain was the ability of atoms to absorb and re-emit certain specific colours of light when energized, for example, the atoms in a neon light. In 1913, the Danish physicist, Neils Bohr (1885–1962), made the daring and intuitive proposal that the electrons in an atom travel around the nucleus only in certain fixed orbits called

Rutherford had predicted that a particle like the neutron could result from the electron falling into the hydrogen nucleus and merging with a proton. In fact, the neutron is an independent particle and is not created by collapsing a hydrogen atom.

energy levels or shells (Figure 22). As long as an electron is moving around the nucleus in one of these energy levels, it does not gain or lose any energy.

An electron is **excited** when it is given additional energy (Figure 23a). The excited electron jumps to a higher energy level further away from the nucleus (Figure 23b).

When the electron loses energy, the electron falls to a lower energy level, closer to the nucleus (Figure 23c). The energy released gives rise to certain colours of light. The colours that we see in fireworks, flames, and neon lights are a result of the energy given out when excited electrons in these materials "relax." (This effect is illustrated in Figure 24.)

Fig. 23

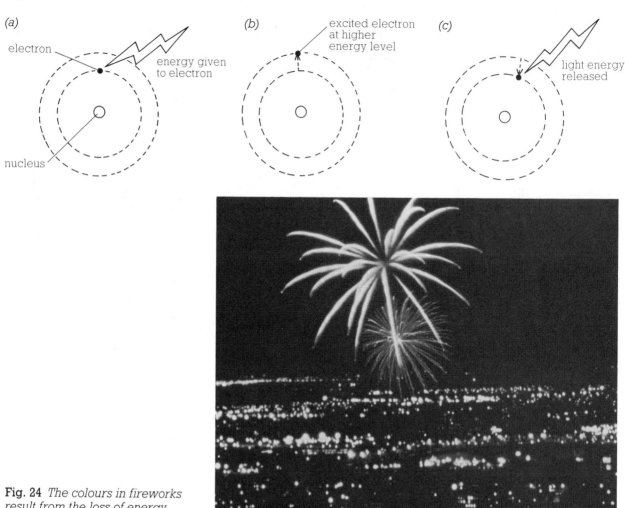

(a)

electron

energy given
to electron

nucleus

(b)

excited electron
at higher
energy level

(c)

light energy
released

Fig. 24 *The colours in fireworks result from the loss of energy when excited electrons "relax."*

Bohr said that the electrons in an "unexcited" atom are in the lowest possible energy levels. Each level, however, can only hold so many electrons before it becomes "full." The following table shows the maximum number of electrons for each level.

Energy level	Maximum number of electrons
1	2
2	8
3	18
4	32

As an example, let us draw a diagram for an atom of chlorine. A chlorine atom has 17 protons in its nucleus and so it must have 17 electrons in the energy levels around that nucleus.

The number of protons in the nucleus of one atom of an element is called the **atomic number** of the element.

(a) Draw a small circle to represent the nucleus (Figure 25a). Inside this nucleus there are 17 protons and a number of neutrons.
(b) Draw a larger circle outside the first circle to represent the first energy level. Place two electrons in this level (Figure 25b). It is now full. There are still 15 more electrons.
(c) Draw a larger circle to represent the second energy level. Place 8 electrons here (Figure 25c). This is the maximum number of electrons for this level. You have accounted for 10 of the 17 electrons.
(d) Draw a larger third level. The maximum number of electrons for this level is 18, but you have only 7 electrons left. Place all 7 electrons here (Figure 25d). You have now accounted for all the electrons.

Such a diagram is called the Bohr-Rutherford diagram because Rutherford proposed the existence of the nucleus and Bohr proposed the ways in which electrons are arranged around the nucleus.

Fig. 25

(a)

chlorine nucleus (17 protons 18 neutrons)

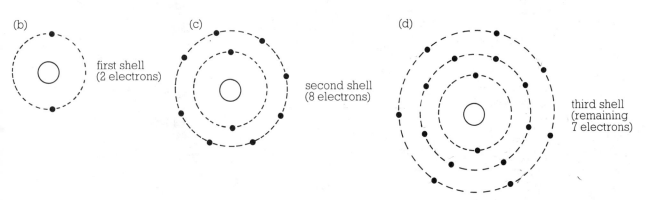

(b)

first shell (2 electrons)

(c)

second shell (8 electrons)

(d)

third shell (remaining 7 electrons)

Questions

1. In what ways does the Bohr-Rutherford atomic model differ from the Rutherford model, and in what ways is it similar?

2. Does the Bohr-Rutherford model meet your criteria for a good model?

3. How could the Bohr-Rutherford model be improved in order for it to explain the most recent findings of science?

4. Draw Bohr-Rutherford diagrams for an atom of each of the following elements. (Each of these atoms also contains neutrons, but the number is not specified here.)
   (a) sodium (11 protons, 11 electrons)
   (b) aluminum (13 protons, 13 electrons)
   (c) phosphorus (15 protons, 15 electrons)
   (d) calcium (20 protons, 20 electrons)

5. The element bromine has an atomic number of 35. For one atom of bromine answer the following questions.
   (a) How many protons are there in the nucleus?
   (b) How many electrons are there outside the nucleus?

P O I N T S · T O · R E C A L L

- Some ancient Greeks, such as Democritus, believed that matter is discontinuous, that is, made up of small, indivisible particles called atoms.
- Aristotle thought matter was continuous, and composed of four elements: fire, air, water, and earth.
- The Dalton model of the atom described it as a solid, indestructible sphere. Dalton regarded atoms of the same element as identical to each other but different from those of other elements, particularly with respect to their masses.
- Static electricity is caused by a transfer of negatively charged particles from one body to another.
- Two objects with like charges (both positive or both negative) repel each other. Unlike charges attract each other.

- Cathode rays are streams of negative particles (electrons) that move toward the anode in a cathode ray tube.
- Becquerel discovered that certain substances are radioactive, that is, they emit high energy radiation capable of exposing film.
- J. J. Thomson proposed a model of the atom in which negatively charged electrons are embedded in a positively charged sphere, like raisins in a plum pudding (or seeds in a watermelon).
- Rutherford determined that radioactive substances can emit three different types of radiation, namely, alpha, beta, and gamma rays.
- In Rutherford's model of the atom, there is a small, positively charged nucleus, which contains most of the mass of the atom. It is surrounded at a relatively great distance by electrons. Most of the atom is empty space.

- The nucleus contains positively charged particles called protons.
- The number of protons must equal the number of electrons in an atom.
- The number of protons in one atom of an element is the atomic number of the element.
- Chadwick discovered the neutron, a subatomic particle with no charge, contained in the nuclei of atoms.

- Bohr modified the Rutherford model of the atom. He said that electrons exist in shells of increasing diameter and increasing energy around the nucleus.
- The maximum number of electrons in the first level is 2, in the second 8, in the third 18, and in the fourth 32.

R E V I E W · Q U E S T I O N S

1. Is the modern view of matter continuous or discontinuous? Explain your answer.

2. How does the Dalton model of the atom explain the fact that compounds are formed by the chemical combination of atoms in whole number ratios?

3. How does the Dalton model of the atom explain the observation that atoms combine in a fixed ratio by mass?

4. Explain why you can pick up little bits of paper with a plastic comb that has been rubbed on your clothing.

5. What properties of the electron can be determined in experiments using cathode ray tubes?

6. Why is the Thomson model of the atom called the "plum pudding model" of the atom?

7. What phenomena can be explained by the Thomson model that cannot be explained by the Dalton model of the atom?

8. (a) What did Rutherford learn about the structure of the atom from his alpha particle scattering experiment?
   (b) How did the results from the experiment enable him to draw those conclusions?

9. Describe Bohr's atomic model and compare it with Rutherford's model.

10. Draw a Bohr-Rutherford diagram to show the arrangement of electrons in the following atoms:
    (a) carbon (6 electrons)
    (b) magnesium (12 electrons)
    (c) silicon (14 electrons)
    (d) sulfur (16 electrons)
    (e) potassium (19 electrons)

11. The element neon has an atomic number of 10. Draw a Bohr-Rutherford diagram for a neon atom.

12. List all models of the atom that can explain each of the following phenomena. If the phenomenon cannot be explained by any of the models you know of, then answer "none."
    (a) Rubbing two materials together may move negative charges from one to the other.
    (b) Oxygen atoms are 16 times heavier than hydrogen atoms.
    (c) Atoms have positively charged nuclei.
    (d) Neon emits a red-orange glow when electricity passes through it.
    (e) Iron is magnetic.
    (f) The charge on a calcium nucleus is twice as high as the charge on a neon nucleus.
    (g) There are five electrons in the outermost (largest occupied) energy level of a nitrogen atom

4 THE PERIODIC TABLE

Helium gas, often used to fill balloons and airships, belongs to a group of unreactive elements. This group is an important link in the periodic arrangement of elements.

CONTENTS

4.1 Introducing the Periodic Table

On the wall of your chemistry classroom, you probably will see a periodic table. This chart displays a method of grouping the elements. Each element occupies a different box on the table. An example of one box, describing the element helium, is shown in Figure 1.

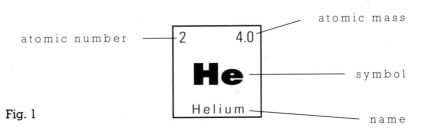

Fig. 1

In the first activity, you will study the first 20 elements in the periodic table; these elements are listed in the following chart.

Element	Symbol	Atomic number	Atomic mass
hydrogen	H	1	1.0
helium	He	2	4.0
lithium	Li	3	6.9
beryllium	Be	4	9.0
boron	B	5	10.8
carbon	C	6	12.0
nitrogen	N	7	14.0
oxygen	O	8	16.0
fluorine	F	9	19.0
neon	Ne	10	20.2
sodium	Na	11	23.0
magnesium	Mg	12	24.3
aluminum	Al	13	27.0
silicon	Si	14	28.1
phosphorus	P	15	31.0
sulfur	S	16	32.1
chlorine	Cl	17	35.5
argon	Ar	18	39.9
potassium	K	19	39.1
calcium	Ca	20	40.1

1 1.0 **H**																	2 4.0 **He**
3 6.9 **Li**	4 9.0 **Be**											5 10.8 **B**	6 12.0 **C**	7 14.0 **N**	8 16.0 **O**	9 19.0 **F**	10 20.2 **Ne**
11 23.0 **Na**	12 24.3 **Mg**											13 27.0 **Al**	14 28.1 **Si**	15 31.0 **P**	16 32.1 **S**	17 35.5 **Cl**	18 39.9 **Ar**
19 39.1 **K**	20 40.1 **Ca**																

Fig. 2 *A simple periodic table.*

Copy the chart into your notebook. For each of the elements, copy its atomic number and its atomic mass from the periodic table (Figure 2). Note that, from one periodic table to another, the position of the atomic number and atomic mass in the boxes may vary. The atomic number is always a whole number.

Do you notice a trend in the atomic numbers? What does the atomic number tell you about the element?

Do you notice a trend in the atomic masses? The trend of the atomic numbers could be described as regular. Is the trend in atomic masses also regular?

The **atomic mass** of an element is the average mass of an atom of that element. The unit of atomic mass is the atomic mass unit (symbol u). The mass of one proton is approximately 1 u; also, the mass of one neutron is approximately 1 u. However, the mass of a single electron is about 2000 times less than 1 u.

By definition, 1 u is equal to $\frac{1}{12}$ of the mass of a carbon atom with 6 protons and 6 neutrons in its nucleus.

Questions

1. How many protons are in the nucleus of one atom of the following?
   (a) carbon   6
   (b) oxygen   8
   (c) sodium   12
   (d) chlorine   18.5

Fig. 3 *Of the three isotopes of hydrogen, only tritium is radioactive.*

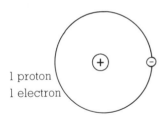

(a)   hydrogen-1

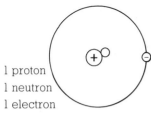

1 proton
1 neutron
1 electron

(b)   hydrogen-2
       (deuterium)

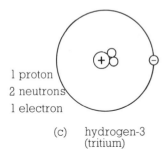

1 proton
2 neutrons
1 electron

(c)   hydrogen-3
       (tritium)

The atomic number of hydrogen is 1.

2. How many electrons are in one atom of the following?
   (a) carbon
   (b) oxygen
   (c) sodium
   (d) chlorine

3. For any element, why is the atomic mass greater than the atomic number?

4. What information do you need to determine the atomic mass of an element from the period table?

5. As you read from left to right in the periodic table, the atomic mass increases. What does this trend indicate about atomic structure?

4.2 Isotopes

If you were introducing a visitor from out of town to your neighbourhood, you might begin by taking your visitor on a walk down your street. You might say, ''Here is where the Berniers live,'' or ''This is the Yamaguchi home next door to the Simpsons'.'' What you have said tells your visiting friend that the people living in each home belong to a different family. Each person in the family is unique, but all members of the family share the same address.

In the periodic table, all hydrogen atoms share the same address, but they are not all identical! Each hydrogen atom has one proton in the nucleus and one electron. It is the number of neutrons that varies from one hydrogen atom to another. Atoms of the same element that have different numbers of neutrons are called **isotopes** of that element. The word ''isotope'' comes from two Greek words — *isos* meaning equal and *topos* meaning place. Since isotopes of the same element have the same address on the periodic table, that is a good name for them!

There are three isotopes of the element hydrogen (Figure 3). These isotopes of hydrogen are very important because of their different physical properties. In the following chart, you will see that each isotope of hydrogen has a different symbol.

Isotope	Symbol	Special name
hydrogen-1	H	hydrogen
hydrogen-2	D	deuterium
hydrogen-3	T	tritium (radioactive)

An isotope of an element is identified by its **mass number**. The mass number is the total number of protons and neutrons found in the nucleus of an atom. The mass number is written after the name of the element as in "hydrogen-1" or "hydrogen-2."

Every element has several isotopes. Some occur naturally like the isotopes of hydrogen and the isotope carbon-14. Carbon-14 is formed in the atmosphere when cosmic rays hit nitrogen-14 atoms. One atom of carbon-14 has 6 protons and 8 neutrons in its nucleus.

Many isotopes that do not occur naturally can be made artificially. The earliest example was phosphorus-30, first prepared in 1934 by Irène and Frédéric Joliot–Curie in France. The usual way of making artificial isotopes is by bombarding atoms with other atoms or parts of atoms. Phosphorus-30, for example, was first made by firing alpha particles, which are helium-4 nuclei, at atoms of aluminum-27.

The equipment needed to make artificial isotopes can be quite complicated. Many such isotopes are now made inside nuclear reactors.

Do not confuse the terms "mass number" and "atomic mass."

Cosmic rays are electrons and the nuclei of atoms, mostly hydrogen, that fall on the earth from outer space.

Questions

1. If you know the mass number and the atomic number of an isotope, then you can easily calculate the number of neutrons in the nucleus of one atom of that isotope with this formula:

   Mass number − atomic number = number of neutrons

   Using this formula, complete the following chart in your notebook. The first calculation has been done for you. Please do not write in your textbook.

Isotope	Atomic number	Number of neutrons
nitrogen-14	7	7
boron-11	5	
aluminum-27	13	
oxygen-18		
lithium-7		
	6	6
	15	16

2. Are two isotopes with the same mass number necessarily isotopes of the same element? Explain your answer.

4.3 Radioactive Isotopes

Working in pairs, study Figures 4 to 9. Write down your positive and negative feelings about the nuclear world we live in. Discuss your feelings and concerns as a class.

When the nucleus of an atom is "unstable," it splits apart, producing other nuclei, particles, rays, and a great deal of heat

Fig. 4 *The ultimate weapon!*

Fig. 5 *A nuclear power plant.*

Fig. 6 *Some foods, such as these strawberries, are preserved by nuclear irradiation.*

Fig. 7 *This patient is being injected with a radioactive solution to help in medical diagnosis.*

Fig. 8 *Cattle in West Germany affected by nuclear fallout from Chernobyl. It is a pity the cow can't read.*

Fig. 9 *Radiation can be used to determine the thickness of metal sheets.*

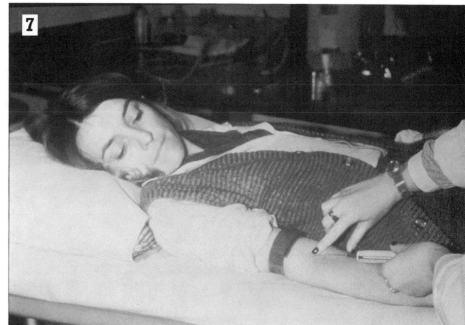

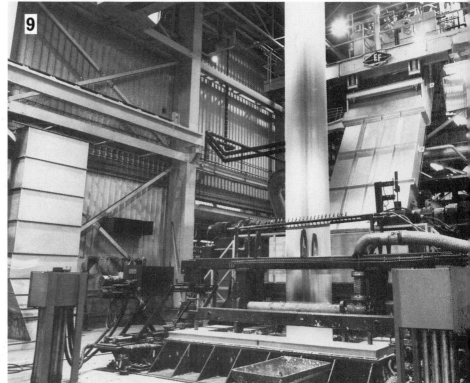

energy. We say that the original nucleus is *radioactive*. The particles and rays given off are called *radiation*. Relatively few radioactive isotopes are found in nature. Except for tritium (hydrogen-3) and carbon-14, most radioactive isotopes are atoms with heavy nuclei. Most of these atoms belong to elements that have atomic numbers greater than 81. A large number of radioactive isotopes that do not exist in nature can be synthesized.

As you learned in section 3.9, the three commonest forms of radiation are denoted by the Greek letters α, β, and γ. The following chart summarizes information about these three forms.

Radiation type	Symbol	Charge	Nature	Penetrating power
alpha, α, particles	^4_2He	2 +	helium nuclei	stopped by one sheet of paper
beta, β, particles	$^0_{-1}e$	1 –	high energy electrons	stopped by 1-cm thick sheet of aluminum
gamma, γ, gamma rays	γ	0	very high energy light waves	stopped by very thick sheets of lead, steel, or concrete

As you can see from the chart, you can protect against radiation by shielding yourself. Gamma rays are quite difficult to stop. Two other ways of protecting yourself are to keep far away from radioactive sources, or, if you are exposed to radiation, to keep your time of exposure as short as possible.

Radiation is highly energetic and may damage the structure of the material it passes through. Cells of living organisms exposed to high levels of radiation may be badly damaged or even die. Low levels of radiation are a natural part of our environment. These low levels of radiation are called **background radiation** and are mainly due to the following:

- cosmic rays from space
- radioactive isotopes produced by cosmic rays
- naturally radioactive isotopes, such as those of uranium and thorium, in the ground and in building materials
- naturally radioactive isotopes in drinking water and food
- radon gas found indoors

Skin cancer, bone cancer, thyroid cancer, leukemia, or other cancers are harmful, sometimes fatal, diseases that may arise from overexposure to radiation. Gamma rays are particularly harmful because they readily pass through the human body.

Questions

1. If radiation is so dangerous, what can you do to shield yourself from it? What makes an effective shield from
   (a) alpha particles?   (b) beta particles?   (c) gamma rays?

2. What else can you do to minimize the harmful effects of exposure to radiation?

3. Using resource materials, make a list of rules that people who work with highly radioactive isotopes every day should follow to protect themselves from overexposure to radiation.

4.4 Beneficial Uses of Radioactive Isotopes

So far we have discussed only radiation that comes from outside of the body. If a radioactive material is taken *into* the body, then the effects can be very dangerous. Gamma rays cause fairly widespread damage in the body; alpha and beta particles can cause considerable damage in a localized area of tissue. Two examples of dangerous isotopes that release damaging radiation are radioactive calcium-45, which deposits along with normal calcium in bones, and radioactive iodine-131, which builds up in the thyroid gland.

Can you think of a way that taking radioactive material into the body can be helpful?

Fig. 10 *The cobalt-60 bomb used to treat cancer is a Canadian invention.*

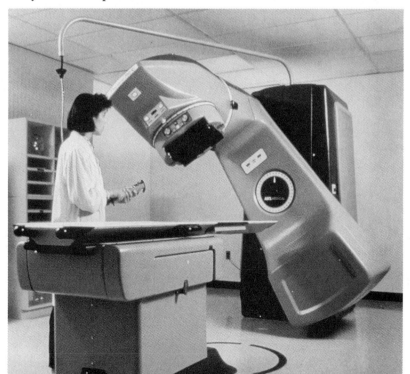

Radioactive isotopes are very useful in medicine. Most hospitals now have nuclear medicine units. The radiation from some radioactive isotopes, such as cobalt-60 and radium-226, is used to destroy cancer cells (Figure 10). Other radioactive isotope help to diagnose internal problems, as shown in the following chart.

Isotope	Use
chromium-51	determining total blood volume
iron-59	measuring how fast red blood cells form and how long they live
arsenic-74	locating brain tumours
iodine-131	detecting and treating defective thyroid glands
phosphorus-32	detecting skin cancer
sodium-24	detecting narrowing or blockage of blood vessels
hydrogen-3 (tritium)	determining total body water

Large pipelines may be blocked from time to time. The blockage can be traced by first putting a radioactive object (tracer) down a pipe. The tracer cannot flow past the blockage and is then stuck at some point below the ground. The point of blockage may be detected above ground using a special device (Figure 11).

Fig. 11 *Radioactive tracers are used to locate blockages in pipelines.*

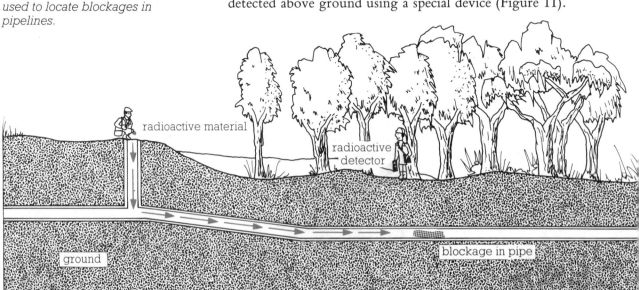

Fig. 12 *Irradiation of potatoes prevents sprouting. Compare untreated potatoes (left) with irradiated potatoes (right) ten weeks after treatment.*

Foods may be irradiated to kill bacteria, delay ripening, and prevent spoilage (Figure 12). The federal government in Ottawa has approved the irradiation of potatoes as a good alternative to the addition of chemical preservatives in foods. Some people are against the irradiation of food. Would you eat food that you knew had been irradiated?

Some gemstones are irradiated to intensify their colour, making them more desirable.

Radiation has applications for insect control. For example, screwworm flies are parasites that infest livestock. The flies are bred in captivity, and the adults are sterilized with radiation. The adult flies are then released to mix with normal flies. Mating results in a high proportion of sterile eggs, so very few flies hatch and the next generation of flies is greatly reduced.

Questions

1. (a) Which radioactive isotope is used to treat cancer of the thyroid gland?
   (b) Why is this isotope chosen?

2. How might radioactive isotopes be used to reduce or prevent an infestation of fruit flies in the crops in your area?

3. Would you wear an irradiated gemstone? Explain your answer.

4.5 Harnessing Nuclear Energy

A controversial use of radioactive isotopes is in the production of nuclear energy. When a nucleus splits, enormous amounts of heat energy are released. The splitting of the nuclei in 1 g of uranium-235 gives about the same quantity of energy as is released when 3 t of coal is burned!

The symbol t stands for tonne.

The production of nuclear energy takes place in nuclear power plants. In a nuclear reactor, the splitting of nuclei, a process known as **nuclear fission**, produces heat that turns water into steam. The steam drives turbines that in turn generate electrical energy (Figure 13). The main parts of a reactor are shown in Figure 14.

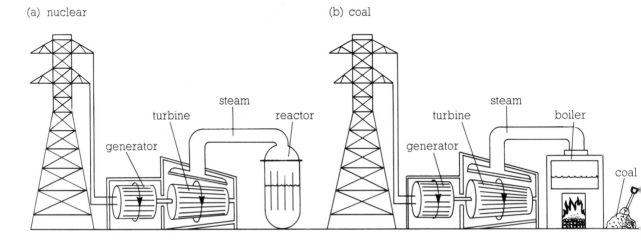

Fig. 13 *A nuclear power plant uses nuclear reactions to produce steam. Other power plants produce steam in other ways.*

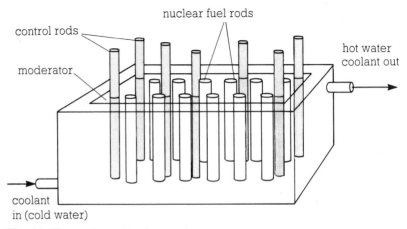

Fig. 14 *The main parts of a reactor.*

Fig. 15 *The Gentilly nuclear reactor.*

A nuclear reactor, such as the one at Gentilly, about 160 km northeast of Montréal, is designed to hold a quantity of nuclear fuel, maintain a ''controlled'' nuclear reaction, and provide for the removal of the heat produced (Figure 15). To control the energy output from a nuclear reaction, only a few nuclei are allowed to undergo splitting at any one time.

Although a reactor uses the same fuel as a nuclear bomb—uranium isotopes—it cannot explode in the same way a nuclear bomb explodes. The reactor is usually protected by thick concrete structures. If the reactor building were to crack open, for example, in an earthquake, huge amounts of radiation would be released and scattered over a large area. The radioactive particles would settle to the ground, causing major hazards to the environment. Soil and water would be contaminated, and the food chain would be affected. Living organisms might have deformed offspring. Livestock and crops might be contaminated by radioactive isotopes, and if so, they would have to be destroyed. Such an event has yet to happen from a natural cause, such as an earthquake, but there have certainly been nuclear accidents, the worst being the one at Chernobyl in 1986.

The settling of radioactive particles is called **nuclear fall-out**.

A large number of radioactive isotopes remain in the used fuel from the reactor. The used fuel is housed in giant cement ''swimming pools'' that are sealed off from other parts of the power plant. The water in the pools absorbs the heat and radiation produced by the radioactive materials. Eventually, long-term storage sites, or safe uses for the spent nuclear fuel, will have to be found because the radiation hazard of the used fuel lasts for hundreds of centuries. Finding suitable locations for waste sites and transporting wastes to the chosen sites may prove to be difficult tasks.

CHEMICAL TIDBITS

VICTIMS OF CHERNOBYL

In April 1986, more than 100 000 Russians were exposed to high levels of radiation when a fire and explosion (not a nuclear explosion) occurred in the nuclear power plant at Chernobyl (Figure 16). This incident was the first well-publicized experience the world has had of an uncontained major nuclear accident. People were evacuated from nearby cities. There were increased radiation levels in Scandinavia within three days of the accident.

As the days went on, doctors in Russia saw patients with vomiting, diarrhea, jaundice, loss of hair, confusion, and high fevers. Some went into a coma and died.

A firefighter who had entered the reactor site suffered radiation burns, which caused his skin to blister and peel. His worst burns were on his hands because he had handled radioactive water. Within ten days he had lost all his hair. His life was saved by a bone marrow transplant carried out by a joint American and Russian surgical team.

Fig. 16 *The destroyed reactor at Chernobyl.*

Questions

1. Does our need for energy warrant the risks associated with nuclear power plants?

2. Imagine an energy crisis. Oil is in very short supply and is being used mainly to make medicine and other essential products. In this situation, nuclear reactors are one of the main sources of energy for power corporations. What would you do if a nuclear reactor site were proposed for an area near your home?

3. People in the area surrounding the Gentilly nuclear facility have complained of a higher than average incidence of cancer. If you were the Minister of the Environment in the province of Québec, what action would you take when you hear these complaints?

4. Nuclear medicine laboratories in hospitals also create nuclear wastes that must be disposed of. What precautions should be taken in transporting these wastes to a disposal site?

5. Choose one of the following topics for a research project:

   - radioactive isotopes in medicine
   - radioactive isotopes in industry
   - radioactive isotopes in energy production
   - radioactive isotopes in food growing or preparation
   - nuclear energy in ships
   - nuclear reactor waste disposal
   - nuclear fall-out: short- and long-term effects
   - a nuclear scientist (Create a mock interview between yourself and the scientist.)

Fig. 17 *The first nuclear-powered vessel, the submarine* Nautilus, *was built in 1954. Now there are other ships driven by nuclear energy.*

You may obtain information on the topic by interviewing people or from sources such as the school resource centre, public libraries, government agencies, pro- and antinuclear agencies. Present your findings to the class in any form *other than an essay.*

6. Since you have a fair knowledge of nuclear matters, you can debate issues such as:

   - nuclear weapons in space
   - nuclear war
   - nuclear dump sites
   - Can countries be totally nuclear free?

Pick one of the above topics. Select two teams: a pro-nuclear team and an antinuclear team. Hold a debate and at the end vote for the team that you think has argued more convincingly.

4.6 Irregularities in the Periodic Table

Having examined the layout of the periodic table and considered the existence of isotopes, let's now see if these two concepts are linked in any way. Let's begin with an atomic scavenger hunt! You may use only a copy of the periodic table and your powers of observation. The task is to find eight cases where the atomic mass *decreases* or *stays the same* as you go from one element to the next in order of increasing atomic number. Who in your class will be the first to find all eight?

As you examined the atomic masses of the elements, you no doubt noticed that most of them are not whole numbers. But how can this be? You know that most of the mass of an atom comes from the masses of the protons and neutrons inside its nucleus. Also, each proton and each neutron has a mass of close to one atomic mass unit (1 u). So, you would expect to get a number that is very close to a whole number when you add the masses of protons and neutrons together. But, for the element chlorine the atomic mass is 35.5 u. How do we explain this number?

You have seen that the atomic masses of elements do not increase regularly as their atomic numbers increase. This fact also seems puzzling. A higher atomic number indicates the presence of more protons in the nucleus of an atom. Each extra proton adds a mass of 1 u. How can it be that the atomic mass sometimes goes down as the atomic number goes up?

If you think that the answers to these questions have something to do with isotopes of the elements, you are on the right track. In nature, the isotopes of an element occur in fairly constant proportions. They are not equally abundant in the universe. Some isotopes are much more common than others.

Among the isotopes of hydrogen, for example, hydrogen-1 (sometimes called protium) is by far the most common. Of all hydrogen atoms in the universe (and that's a lot of atoms because hydrogen is the most abundant element in the universe!), fully 99.98% of them are protium atoms. Only 0.015% are deuterium atoms. Tritium is rarer still.

If you were to calculate the average mass of a student in your class, what would that information mean? Would it mean that everyone in your class now has the same mass? Is there even one person in the class whose mass exactly matches that of the "average student"? This example of average mass is similar to the atomic mass of an element. The atomic mass is an average. All atoms of the same element do not have the same mass. Some isotopes have more neutrons and so have a higher mass than others. Perhaps there is no isotope whose mass exactly matches the average atomic mass.

Fig. 18 *The symbols in this illustration show another way of representing isotopes. The atomic and mass numbers are shown to the left of the element symbol.*

The mass number of an isotope tells you the mass, in u, of one atom of that isotope.

Scientists calculate the "average" mass of the naturally occurring isotopes of an element to determine the atomic mass of that element. Chlorine, for example, has two naturally occurring isotopes, chlorine-35 and chlorine-37 (Figure 18). In a natural sample of chlorine, 75.0% of the atoms are chlorine-35 atoms, each with a mass of 35.0 u. The remaining 25.0% of the atoms are chlorine-37 atoms, each with a mass of 37.0 u.

Example 1

Calculate the atomic mass of chlorine from the preceding information.

Solution

The atomic mass of chlorine is the average mass of naturally occurring chlorine atoms. We know the following information:

Isotope	Abundance (%)	Mass (u)
chlorine-35	75.0	35.0
chlorine-37	25.0	37.0

Since the abundance of each isotope is given as a percentage, consider 100 atoms of chlorine. Out of every 100 atoms, 75 are chlorine-35 atoms and 25 are chlorine-37 atoms.

Mass of 75 chlorine-35 atoms = 75 × 35.0 = 2625 u
Mass of 25 chlorine-37 atoms = 25 × 37.0 = 925 u
Total mass of 100 chlorine atoms = 3550 u

$$\text{Average mass of 1 chlorine atom} = \frac{3550\,u}{100}$$
$$= 35.5\,u$$

Compare this answer with the atomic mass of chlorine in the periodic table.

Notice there is no naturally occurring isotope of chlorine with a mass number that equals the atomic mass of chlorine.

The chemical behaviour of isotopes of an atom depends only on the number of electrons. Since atoms of all isotopes of an element have the same number of electrons, all isotopes of an element have the same chemical properties.

Questions

1. Is it possible that the atoms of an element have different atomic numbers? Explain.

2. Is it possible that the atoms of an element have different mass numbers? Explain.

3. The atomic mass of magnesium is not a whole number. Why not?

4. One-half of all the bromine atoms in the universe have a mass number of 79 and the other half have a mass number of 81. What is the average mass of a bromine atom? What is the atomic mass of bromine?

5. How do you account for the fact that no bromine atom in the universe has a mass number equal to the atomic mass of the element?

6. Eighty percent of all boron atoms have a mass number of 11. The rest have a mass number of 10. What is the atomic mass of boron?

7. Explain the irregularities you discovered in atomic masses of the elements in the periodic table.

8. Some elements, like neptunium (number 93), have atomic mass values that do not look like most other values. The atomic mass of neptunium is a whole number, 237. There is no figure after the decimal point, not even a zero. Find out why the atomic masses of neptunium and some other elements are written in this way.

9. There are two naturally occurring isotopes of lithium, lithium-6, and lithium-7. Which isotope is more abundant in nature?

| Investigation | 4.7 Sorting Out the Elements |

You now have a sense that the arrangement of the periodic table is connected with the order of atomic numbers, and to a lesser extent, with the order of average atomic masses. But what about the characteristic properties of the elements? Is there any link between the properties of elements and the way elements are arranged in the periodic table? Let's take a look at some elements and explore their characteristic properties.

Materials

safety goggles	test tubes
mortar and pestle	test tube rack
emery paper	dilute hydrochloric acid
conductivity apparatus	

elements: aluminum, iron, copper, tungsten, magnesium, carbon (graphite), silicon, sulfur, oxygen, nitrogen, neon

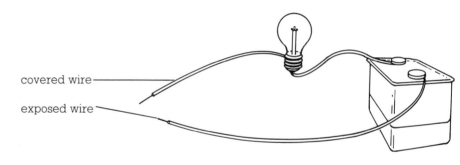

Fig. 19 *A simple conductivity apparatus.*

Method

1. Prepare a data chart in your notebook, such as the one below. Enter your observations as you go along.

| Element | Description | Malleable? | Conducts | | Reacts with acid? |
			heat?	electricity?	

2. Describe the appearance of each element. Record such features as state, colour, and shininess.

Malleable means capable of being hammered into different shapes without shattering.

3. One at a time, put a sample of each of the solid elements into the mortar and mash it gently with the pestle. Do any of the elements "flatten out" a bit? Do some of them crumble? Decide which of the elements you could hammer into a sheet without breaking it into pieces.

4. Rub each solid element with emery paper. Does the element feel warmer after you do this? Which elements are good conductors of the heat of friction?

Do not touch the wires with your fingers!

5. Let the wire ends of the conductivity apparatus make contact with a sample of each element. Does the bulb light up when you connect with certain elements? Which ones are they?

6. Put a small sample of each solid element into a different test tube. Add a few drops of dilute hydrochloric acid to each test tube. Describe what happens. Which elements react with the hydrochloric acid and which do not?

Follow-up

1. Based on your data table, are there categories of elements with similar general properties? How many categories are there? Which elements fall into each category?

2. What is the symbol for each element?

3. Taking one category at a time, locate the elements on an outline of the periodic table (Figure 20). Please do not write in your textbook.

4. The elements that occur to the *left* of the heavy, step-like line on the periodic table are known as **metals**. What properties do these elements have in common?

Fig. 20 *A version of the modern periodic table.*

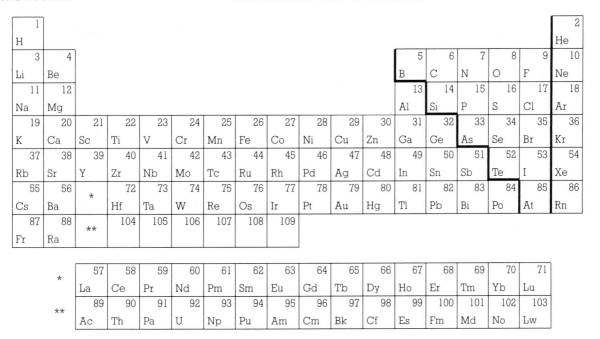

Fig. 21 *The division into metals and nonmetals in the periodic table.*

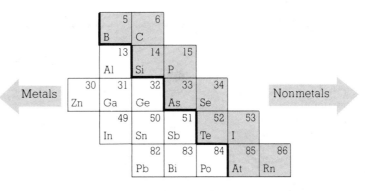

5. The elements that lie to the *right* of the heavy, step-like line are **nonmetals.** What properties do these elements have in common?

6. Did you find any elements that had some properties of metals and also some properties of nonmetals? For these elements that "sit on the fence," choose the category of elements they are most similar to. Why did you make that choice? We can actually make a new category for these in-between elements. They are are called **metalloids** or **semimetals**.

7. Which of the following elements are metals, which are non-metals, and which are metalloids?

arsenic (As)	tin (Sn)	bismuth (Bi)
gold (Au)	tungsten (W)	radium (Ra)
radon (Rn)	chlorine (Cl)	hydrogen (H)
mercury (Hg)	lead (Pb)	iodine (I)
germanium (Ge)	cobalt (Co)	antimony (Sb)

8. What is the meaning of the heavy, step-like line on the periodic table?

9. Which of the following elements should be good conductors of electricity? Explain.
   (a) silver (Ag)　　　　　(c) uranium (U)
   (b) iodine (I)　　　　　 (d) xenon (Xe)

10. Which of the following solid elements should crumble when hit? Explain.
   (a) zinc (Zn)　　　　　 (c) calcium (Ca)
   (b) iodine (I)　　　　　 (d) gold (Au)

Investigation	*4.8 Getting to Know Some Families*

Do not touch the metals used in this experiment because some of them react with moisture. Use a large sheet of clean paper to transfer small pieces of the metals. Wear safety goggles at all times.

While all the elements that we categorized as metals were shiny and were excellent conductors of heat and electricity, they certainly were not all exactly the same. Which metal was a different colour? which was the softest? which reacted most vigorously with the acid? All metals are not alike and neither are all nonmetals. Let's see if we can break up these large categories into smaller ones. In this investigation, you will be looking at the reactions of several metals and a few nonmetals.

Fig. 22 *Chlorine, a poisonous yellowish-green gas, has many industrial uses. For example, it is used to kill bacteria in swimming pools. It was also used in World War I as a deadly weapon.*

Chlorine water and bromine water are the common names for solutions of the elements chlorine and bromine in water.

A **tincture** is a solution of a substance in alcohol.

Materials

safety goggles
250-mL beaker
test tubes
test tube rack
ceramic gauge pad
glass rod
cyclohexane
phenolphthalein solution
sodium iodide solution
chlorine water
bromine water
tincture of iodine
metals: lithium, calcium, magnesium

Fig. 23

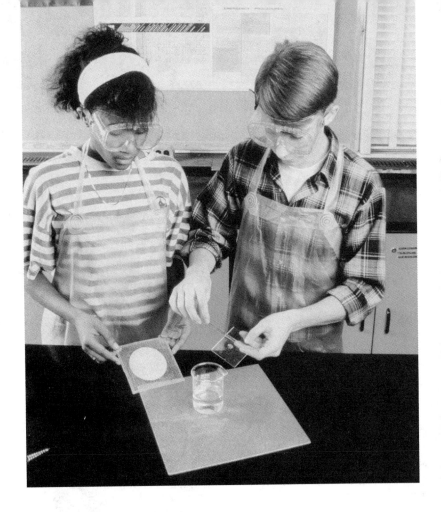

The piece of lithium should be no bigger than 1 mm³.

Method

1. Draw up a suitable chart to record all your observations.

2. Use a dry glass rod to add a small piece of one of the metals to about 150 mL of fresh water in a clean 250-mL beaker (Figure 23). Immediately cover the beaker with the ceramic gauze and observe from a distance of about 1 m.

3. When the reaction, if any, dies down, add a drop of phenolphthalein to each solution and observe any colour change.

4. Repeat steps 2 and 3 for each of the other metals. Use a small piece of metal each time and a fresh sample of water.

5. Place about 3 mL of sodium iodide solution in a test tube.

6. Add about 1 mL of chlorine water to the test tube and mix well. Then add about 1 mL of cyclohexane and mix well. Observe any colour changes.

7. Repeat steps 5 and 6, but use bromine water instead of chlorine water.

8. Pour about 3 mL of tincture of iodine into a test tube. Add 1 mL of cyclohexane and mix well. Observe the colour change.

Follow-up

1. (a) What happened when the lithium and calcium reacted with water?
   (b) What happened when the phenolphthalein was added?
   (c) What evidence suggests that these metals react with water in the same way?
   (d) Did both react with water at the same speed?

2. A tiny piece of potassium reacts explosively with water. Sodium does not react quite as quickly as potassium but faster than lithium. Arrange lithium, sodium, and potassium in order of *increasing* reactivity, from slowest to fastest.

3. Based on your observations for magnesium and calcium, place these metals in order of increasing reactivity.

4. From your observations and from the information provided in question 2, compare potassium to calcium and sodium to magnesium. Which metals are more reactive?

5. Locate in the periodic table the five metals mentioned in questions 2 and 3.

6. (a) What happened when you added chlorine water to the sodium iodide solution?
   (b) What happened when you added bromine water?
   (c) How do the colours of the resulting solutions compare to the colour of the tincture of iodine?

7. (a) What happened when cyclohexane was added to the tincture of iodine?
   (b) What do you think happened to cause the colour to move from the alcohol solution into the cyclohexane?

8. Based on your observation, what formed when either chlorine water or bromine water was added to sodium iodide solution?

9. Which reaction was faster? Based on your observations, list bromine, chlorine, and iodine in order of increasing reactivity.

10. Locate the elements bromine, chlorine, and iodine in the periodic table.

4.9 The Modern Periodic Table

There are seven periods. To make the width of the periodic table more manageable, Periods 6 and 7 are each shortened by removing blocks of elements. These blocks are shown separately below the main table.

The modern periodic table (Figure 24) is arranged in horizontal rows (called **periods**) and vertical columns (called **groups** or **families**). Elements within a group are often referred to as a chemical family of elements. The members of a family have similar chemical properties. The eight longest columns (two on the left-hand side and six on the right-hand side) are called Groups 1 to 8. The elements in these groups are frequently called the **representative** or **main group** elements. You sometimes hear these groups referred to as the "A" groups. The group numbers are sometimes written as IA-VIIIA, rather than just 1-8. Both systems are shown on the more detailed periodic table at the back of this book.

The word "alkali" comes from the Arabic word for ashes (*al-qali*) of certain plants. These ashes were known by the Arabic alchemists to have unique properties in water.

The elements of Group 1 are often referred to as the **alkali metals**. They are so reactive that they are never found as free elements in nature, only as parts of compounds.

Fig. 24 *The modern periodic table.*

Period	Group 1	2	transition metals										Group 3	4	5	6	7	8
1	1 H																	2 He
2	3 Li	4 Be											5 B	6 C	7 N	8 O	9 F	10 Ne
3	11 Na	12 Mg											13 Al	14 Si	15 P	16 S	17 Cl	18 Ar
4	19 K	20 Ca	21 Sc	22 Ti	23 V	24 Cr	25 Mn	26 Fe	27 Co	28 Ni	29 Cu	30 Zn	31 Ga	32 Ge	33 As	34 Se	35 Br	36 Kr
5	37 Rb	38 Sr	39 Y	40 Zr	41 Nb	42 Mo	43 Tc	44 Ru	45 Rh	46 Pd	47 Ag	48 Cd	49 In	50 Sn	51 Sb	52 Te	53 I	54 Xe
6	55 Cs	56 Ba	*	72 Hf	73 Ta	74 W	75 Re	76 Os	77 Ir	78 Pt	79 Au	80 Hg	81 Tl	82 Pb	83 Bi	84 Po	85 At	86 Rn
7	87 Fr	88 Ra	**	104	105	106	107	108	109									

*	57 La	58 Ce	59 Pr	60 Nd	61 Pm	62 Sm	63 Eu	64 Gd	65 Tb	66 Dy	67 Ho	68 Er	69 Tm	70 Yb	71 Lu
**	89 Ac	90 Th	91 Pa	92 U	93 Np	94 Pu	95 Am	96 Cm	97 Bk	98 Cf	99 Es	100 Fm	101 Md	102 No	103 Lw

Their "neighbours" on the periodic table, the elements of Group 2 are known as the **alkaline earth metals**. They are almost as reactive as the alkali metals and are also not found free in nature. The "earth" in the name refers to the fact that these metals are abundant on this planet. Calcium is the fifth most abundant element in the earth's crust and magnesium is the sixth.

The nonmetals in Group 7 are called the **halogens**. These coloured, reactive elements are never found free in nature.

All the elements in the middle of the table, between Groups 2 and 3, are called **transition metals**.

The word "halogen" is derived from the Greek words *hals*, meaning salt, and *genes*, meaning born. They are elements that form salts. The most familiar salt, sodium chloride, is formed from the elements sodium (an alkali metal) and chlorine (a halogen).

Fig. 25 *Hydrogen, used as a fuel for the space shuttle, is not a metal. You will learn later why hydrogen is placed in group 1 even though its chemical properties are different from those of the other group 1 elements.*

Chemically Speaking

MORE ABOUT THE ELEMENTS

The known elements are not distributed evenly throughout the universe. Hydrogen is the most abundant element in the universe—about 90% of all atoms are hydrogen atoms. Helium atoms make up most of the rest.

Twelve elements were known before the sixteenth century: antimony, arsenic, bismuth, carbon, copper, gold, iron, lead, mercury, silver, sulfur, and tin.

Some elements are named after people, for example, curium (Curie), and einsteinium (Einstein). Other elements are named after places, such as californium (California) and polonium (Poland). Four elements are named after Ytterby, the town in Sweden where they were discovered. The four elements are erbium, terbium, ytterbium, and yttrium.

It is thought that the composition of the elements of the universe is constantly changing. Hydrogen is being changed into helium inside stars (like our sun), and helium is being changed into the other elements.

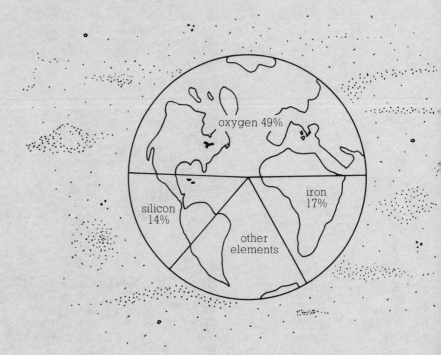

Fig. 26 *The most abundant elements on earth.*

oxygen 49%

iron 17%

silicon 14%

other elements

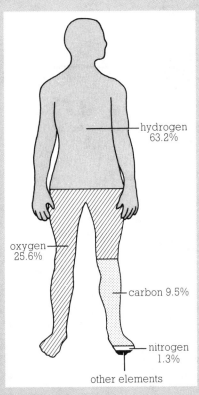

hydrogen
63.2%

oxygen
25.6%

carbon 9.5%

nitrogen
1.3%

other elements

Fig. 27 *The elements making up the human body.*

About 70 elements have been identified in the sun. The element helium was discovered in the sun before it was discovered on earth. In fact, the name "helium" comes from the Greek word *helios*, meaning the sun.

The most abundant element found on the earth itself is oxygen (about 49% of all the atoms), followed by iron (17%), and then silicon (14%). These three elements, together with magnesium, sulfur, nickel, aluminum, and calcium, make up about 99% of all the atoms found on earth.

In the human body, 63.2% of the atoms are hydrogen, 25.6% oxygen, 9.5% carbon, 1.3% nitrogen, 0.2% phosphorus, and 0.2% all the rest.

By international agreement, each element has a unique symbol. It does not matter what language is spoken or what method of writing is used, everyone uses the same symbols for the elements. Every symbol consists of one or two letters. The first, and in some cases only, letter is a capital letter.

Many of the symbols seem obvious when you consider the names of the elements. Some symbols, however, bear no resemblance to their corresponding names. The symbol, W, used for the element tungsten, comes from the German name *wolfram*. The symbols of other elements are related to their Latin names. Your knowledge of French will help you remember some of them.

Element	Latin name	French name	Symbol
copper	cuprum	cuivre	Cu
gold	aurum	or	Au
iron	ferrum	fer	Fe
lead	plumbum	plomb	Pb
mercury	hydrargyrum	mercure	Hg
potassium	kalium	potassium	K
silver	argentum	argent	Ag
sodium	natrium	sodium	Na
tin	stannum	étain	Sn

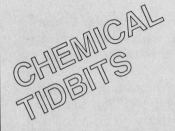

THE NOBLEST ELEMENTS OF THEM ALL

The last group of elements to be discovered was Group 8 — the noble gases, helium through to radon. These elements did not at first appear to react with any other element. However, in 1962, Neil Bartlett (Figure 28) at the University of British Columbia formed the first compound of xenon. Today, other xenon compounds and some krypton compounds are known.

Fig. 28 *Neil Bartlett showed in 1962 at the University of British Columbia that the noble gas xenon could react to form a compound. In the photograph he is holding the apparatus in which the original reaction took place.*

Questions

1. Recall the reactivity of the five metals you discussed in investigation 4.8. Do the metals become more reactive
   (a) as you go up or down a group?
   (b) as you move to the right or to the left across a period?
   (c) On the basis of your answers to (a) and (b), predict the identity of the most reactive metal in the periodic table.

2. Recall the reactivity of the three nonmetals you studied in investigation 4.8.
   (a) Do they become more reactive as you go up or down the group?
   (b) How does the trend for the nonmetals compare with the trend you observed for a group of metals?
   (c) Predict whether nonmetals become more reactive as you move to the right or to the left across a period.
   (d) Predict the most reactive nonmetal in the periodic table. (Remember that helium is a noble gas, so it is extremely unreactive.)

3. Draw an outline of the periodic table. On your diagram, indicate where the noble gases are located. Show the dividing line between metals and nonmetals. Summarize your findings from investigation 4.8 by showing on the diagram each of the following items.
   (a) the location of the metals
   (b) the location of the nonmetals
   (c) the general directions, both horizontal and vertical, in which metals become more reactive
   (d) the general directions, both horizontal and vertical, in which nonmetals become more reactive

4. Based on the results from investigation 4.8, select the more reactive element from each pair of elements following:

Li or Be	Zn or Al	F or S	He or F	C or Si
Rb or Ca	Pt or Au	Pb or Sn	Sr or Sc	As or Se

4.10 Predicting Properties From Periodic Trends

The circles that you see in Figure 29 represent the relative sizes of the atoms of elements with atomic numbers from 1 to 20, excluding the noble gases.

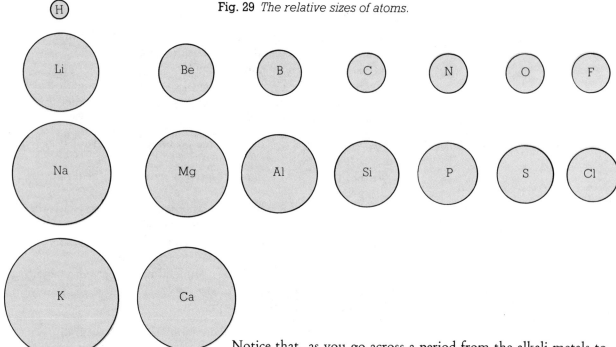

Fig. 29 *The relative sizes of atoms.*

Keep in mind the organization of electrons in an atom!

Notice that, as you go across a period from the alkali metals to the halogens, the atoms within a period are approximately the same size. This is true for each of the periods. In groups, suggest a reason for this trend.

If you are really sharp, you may have noticed that, going across a row, the size of the atom actually shrinks a bit. The slight contraction is caused by the increased charge and pull of the positive nucleus on the negative electrons.

Working with classmates, prepare a table of the first 36 elements, their symbols, and their atomic numbers. Select two properties from among the following: density, melting point, boiling point, and electrical conductivity. Using reference resources, record the values of your two chosen properties for each of the first 36 elements.

Construct two large graphs, one for each property, with the units of the property along the y-axis and the atomic numbers of the elements along the x-axis. On each graph, locate the elements of period 1, period 2, period 3, and period 4. Share your group's graphs with the rest of the class.

What is the trend for each of your chosen properties within a given period? Does the property's value stay the same across the period? Does it increase? Does it decrease?

Summarize your observations of periodic trends in a few sentences and record them in your notebook.

Questions

1. Scientists have estimated the size of the atoms of different elements. The bar graph in Figure 30 shows the atomic radii of the first 20 elements in the periodic table. The values for two of the elements are missing.
   (a) What evidence suggests that the atomic radii of the elements follow a periodic trend?
   (b) Use the graph to estimate the atomic radius for each of the two missing elements.

$1 \text{ pm} = 1 \times 10^{-12} \text{ m}$

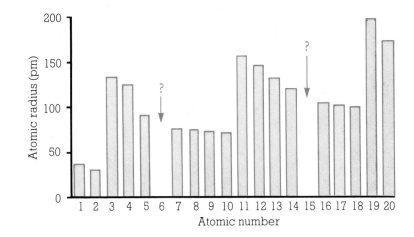

Fig. 30 *The atomic radii of the first 20 elements.*

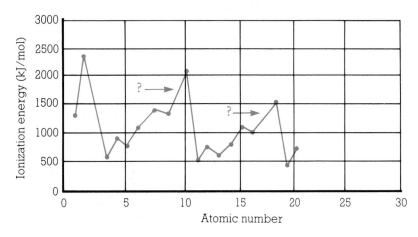

Fig. 31 *The ionization energies of the first 20 elements.*

2. The term **ionization energy** is used to describe how much energy is needed to completely remove an electron from an electrically neutral atom. The line graph in Figure 31 shows the values of the ionization energies for the first 20 elements in the

periodic table. The values for two of the elements are missing.

(a) What evidence suggests that ionization energy values for the elements form a periodic trend?

(b) Does the ionization energy generally increase or decrease across a period?

(c) Use the graph to estimate the ionization energy of each of the two missing elements.

3. The bar graphs in Figures 30 and 31 are not the only way of representing the same numerical data. Figures 32 and 33 show plots on coordinate axes for a larger number of elements. Do these graphs indicate that the periodic trends you found in questions 1 and 2 continue throughout the table? Explain.

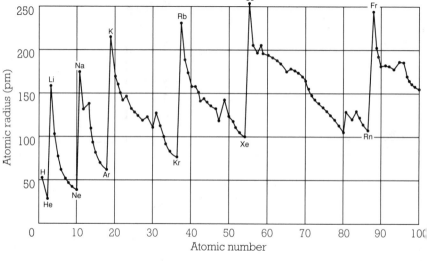

Fig. 32 *The atomic radii of a large number of elements. Note that individual values do not agree with those in Figure 30, as a different source of data was used.*

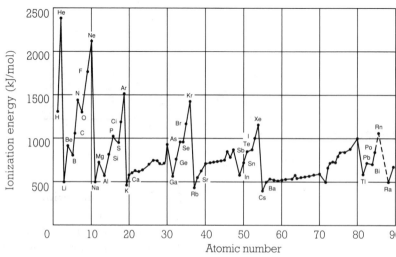

Fig. 33 *The ionization energies of a large number of elements.*

4.11 Technological Uses for the Elements

Here is a matching-columns puzzle for you to solve. In the left column are the symbols of some elements. On the right is a list of technological uses for these elements. Your task is to match each use with the correct element. Some are easy to match and you will get them right away. Others may take some research on your part. See how many you can match.

Please do not write into your textbook. Copy the list of element symbols in your notebook and write the number of the matching use beside each symbol.

Symbol		Use
W	( )	1. It is one of the elements mixed with mercury to make tooth fillings, not only because it is an attractive element, but also because it has antiseptic properties.
U	( )	2. Computer chips are made of this element.
Se	( )	3. In the form of diamond, this element is used to cut through hard materials; as graphite, it is a lubricant and conducts electricity.
He	( )	4. It fuels nuclear reactors.
H	( )	5. Aircraft manufacturers make lightweight alloys from this metal.
C	( )	6. This element is used in matchheads because it is easily ignited.
Pt	( )	7. It is used in the process which converts liquid vegetable oils into solid fats, such as margarine.
P	( )	8. Because it supports combustion, it is mixed with flammable gases like acetylene in welding torches.
Si	( )	9. "Silent" electric switches contain this element—so do barometers!
Cu	( )	10. This metal is found in the filaments of light bulbs.
O	( )	11. This precious metal with a very high melting point does not corrode. Fibreglass is made by forcing molten glass through tiny holes in a sheet of this metal.
Ag	( )	12. When you photocopy something, light creates an image on a thin film of this metalloid. "Toner" sticks to the sensitized areas of the film and reproduces the image on a sheet of paper.
Al	( )	13. Because it cannot burn and is lighter than air, it is used to fill airships and balloons.
Hg	( )	14. Because of its high electrical conductivity, it is the metal most often used for electrical wiring.

4.12 A Family Portrait

Where does the third electron go when it is added to an atom? the eleventh electron?

If you take another look at the atomic radii illustrated in Figure 30, you should notice how the size of the atoms in a group or family grows larger as the atomic number increases. In groups, propose an explanation for this observation.

Take another look at the graphs you prepared in section 4.10 and those in the questions at the end of that section. Locate the alkali metal family (Group 1) on each of the graphs. How does the property which you have plotted against the atomic number change as you go from one element in the family to the next? Does the same thing happen in the alkaline earth family? In the halogen family? In each family?

In a sentence or two, summarize your observations of the trends within a family or group on the periodic table and write it in your notebook.

Questions

1. Examine the graph of the ionization energies of the first 20 elements (Figure 31).
   (a) What is the trend in ionization energies of the elements in a group?
   (b) Compare the trend in ionization energies with the trend in atomic radii for elements in the same group.

2. In section 4.11, you matched the symbol for an element with its appropriate use. The elements represented various parts of the periodic table. Do likewise for the following elements drawn from the alkali metals, alkaline earth metals, and halogens.

Symbol		Use
Ba	( )	1. Street lamps that give off a bright yellow light contain this metal as a vapour.
Be	( )	2. Found in potash, this element promotes growth in plants. It is an important component in fertilizers.
Na	( )	3. The carbonate salt of this metal is used in the treatment of manic depression, although the way in which it is able to relieve the extensive mood swings associated with this ailment is unknown.
F	( )	4. This reactive alkaline earth metal is used to remove traces of oxygen and nitrogen gas from vacuum tubes.
Li	( )	5. When this element is mixed with copper it produces a long-wearing metal that does not spark when struck. For this reason, it is the material of choice for tools used to work with explosive gases.

Symbol	Use
Cl ( )	6. When alloyed with aluminum, this metal produces a strong, light alloy useful for lightweight tools and aircraft parts. Perhaps the trim on your luggage is made of such an alloy.
Mg ( )	7. Teflon is manufactured from this highly reactive halogen.
I ( )	8. Most of this element which is produced from the electrolysis of salt water, is used in the manufacture of PVC (polyvinylchloride) plastics.
K ( )	9. A good deal of this element is used in producing the silver halide salts needed for high-speed photographic film.

4.13 The Model of the Atom and Periodic Behaviour

There are hardly more varied elements in appearance than the members of the halogen family. Two are pale yellow gases (chlorine and fluorine), one a fuming reddish-brown liquid (bromine), and another a purplish-black crystal (iodine). However, why are these these elements so reactive and why do they behave in similar ways? It must be something their atoms have in common.

In your notebook, make a large copy of the simplified periodic table in Figure 34. In the boxes on the periodic table, draw Bohr-Rutherford diagrams of the first 18 elements.

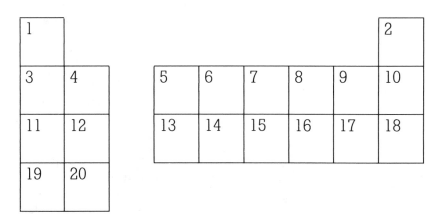

Fig. 34 *A simplified periodic table.*

The electrons in the outermost occupied energy level of an atom are known as its **valence electrons**.

Questions

1. For the halogens fluorine and chlorine, in what way are their arrangements of electrons
   (a) similar?
   (b) different?

2. How many electrons are in the outermost occupied energy level of fluorine and chlorine atoms?

3. (a) How does your answer to question 2 relate to the number of the group occupied by fluorine and chlorine on the periodic table?
   (b) In your notebook, put the group number on the top of the column of the periodic table that contains the halogen family.

4. Add a box to your simplified periodic table for the element bromine. Draw the Bohr-Rutherford diagram for bromine. How does the number of electrons in the outermost occupied energy level in bromine compare with the halogen group number and your answer to question 2?

5. Without drawing the Bohr-Rutherford diagram for iodine, predict the number of electrons in its outermost occupied energy level.

6. (a) Compare the outermost occupied energy levels of the elements in Group 1. What do they have in common?
   (b) What do the elements in Group 2 have in common?
   (c) Label the columns containing the alkali metals and the alkaline earth metals with their respective group numbers.

7. Compare the number of electrons in the outermost occupied energy levels of the elements in Groups 3, 4, 5, and 6 with their respective group numbers. Do these groups fit the same pattern that you have already observed?

8. (a) Does the pattern you have observed continue for the noble gases (Group 8)?
   (b) Why is helium an exception?
   (c) In what way does helium's outermost occupied energy level resemble those of neon and argon?

Any group of eight is called an "octet." An octet of electrons in the outermost occupied shell seems to be a "magic number" for elements.

9. If the outermost occupied shells of an element's atoms are filled, then the elements seems to be unreactive. Such elements are said to be **stable** because they remain unchanged. In which group do these types of elements belong?

10. How many occupied shells are there in
    (a) a fluorine atom?
    (b) a chlorine atom?

11. How does the number of occupied shells compare in atoms of
    (a) fluorine, lithium, and beryllium?
    (b) chlorine, sodium, and magnesium?

12. How does the number of occupied shells in an atom relate to the number of the period in which it is located, starting with hydrogen and helium as period 1? Based on this conclusion, label all the periods on your periodic table.

13. Without drawing Bohr-Rutherford diagrams, predict the number of valence electrons in one atom of each of the following elements.
    (a) Sn   (b) At   (c) Xe   (d) Cs   (e) Ba

14. How is the Bohr-Rutherford model of the atom useful in predicting the chemical behaviour of an element?

15. In periodic tables that number the main or representative groups as IA-VIIIA, there is a second set of groups numbered IB-VIIIB. The elements in these groups are the transition metals. The first row of the transition metals includes elements 21 to 30. You have already seen that the properties of elements in the "A" groups can be radically different from one group to another. For the "B" groups, properties are generally quite similar.
    (a) Draw Bohr-Rutherford diagrams for the following elements.
        (i) Sc   (ii) Mn   (iii) Ni   (iv) Zn
    (b) Do your diagrams in (a) indicate why the B groups are numbered in the way described above?
    (c) Do your diagrams in (a) indicate why these four transition metals have fairly similar properties?

16. (a) Draw Bohr-Rutherford diagrams for the following:
        (i) potassium   (ii) calcium
    (b) Does each diagram in (a) predict the expected number of valence electrons for each element?

17. In the light of your answers to questions 15 and 16, re-assess the Bohr-Rutherford model of the atom. What are the model's strengths and weaknesses?

4.14 The Development of the Periodic Table

Fig. 35 *Dmitri Mendeleev's theory of periodic trends in the properties of elements (1872) is the basis of our modern periodic table.*

				Ti = 50	Zr = 90	? = 180.
				V = 51	Nb = 94	Ta = 182.
				Cr = 52	Mo = 96	W = 186.
				Mn = 55	Rh = 104,4	Pt = 197,4
				Fe = 56	Ru = 104,4	Ir = 198.
			Ni = Co = 59		Pl = 106₆,	Os = 199.
H = 1				Cu = 63,4	Ag = 108	Hg = 200.
	Be = 9,4	Mg = 24	Zn = 65,2		Cd = 112	
	B = 11	Al = 27,4	? = 68		Ur = 116	Au = 197?
	C = 12	Si = 28	? = 70		Sn = 118	
	N = 14	P = 31	As = 75		Sb = 122	Bi = 210
	O = 16	S = 32	Se = 79,4		Te = 128?	
	F = 19	Cl = 35,5	Br = 80		I = 127	
Li = 7	Na = 23	K = 39	Rb = 85,4		Cs = 138	Tl = 204
		Ca = 40	Sr = 87,4		Ba = 137	Pb = 207.
		? = 45	Ce = 92			
		?Er = 56	La = 94			
		?Yt = 60	Di = 95			
		?In = 75,6	Th = 118?			

Several chemists around 1860 proposed that the properties of elements followed periodic trends. John Newlands from England, Dmitri Mendeleev from Russia, and Lothar Meyer from Germany were some of these chemists. Only Mendeleev is now widely remembered for this idea.

Around 1860, scientists were quite preoccupied with the 60 or so elements that were then known. They were attempting to further purify the elements so that accurate measurements could be made of their properties. Many of the scientists were looking for some sort of pattern in the properties of these elements that would simplify the study of chemistry.

In 1872, Mendeleev (Figure 35) published a chart of the then known elements. In this chart the elements were arranged in order of their atomic masses. Elements with similar properties were grouped together. Because some of the properties were seen to repeat on a regular or periodic basis, this chart became known as a periodic table of the elements. A great deal of the data of those days were inaccurate and this led to some inconsistencies in Mendeleev's periodic table. Mendeleev, however, was so convinced he was correct that he used his periodic table to predict which of the elements had incorrect data. He also predicted properties of elements that had not yet been discovered. He did this by using trends in the properties of known elements.

At first, few scientists agreed with Mendeleev's idea. When the elements Mendeleev had predicted were later discovered and shown to have properties close to those predicted, scientists became more convinced of the usefulness of his periodic table.

Mendeleev's arrangement of the elements according to their atomic masses led to some inconsistencies. After the discovery of protons, neutrons, and electrons, the order of some of the elements was changed. Scientists realized that an atom of each element is distinguished by the number of protons in its nucleus. In today's version of the periodic table, the elements are arranged in order of *increasing atomic number*, not atomic mass.

Questions

1. At the time of Mendeleev, the element uranium was thought to have an atomic mass of about 116 u. Today we believe the correct value should be about 238 u. Why do you think these values are so different?

2. One whole family of elements was not known at the time of Mendeleev. The first member of this group of elements was found on the sun. This was done by viewing the sun during a solar eclipse and analyzing the light given off. The analysis suggested the presence of an unknown element. The element was named helium after the Greek word *helios*, meaning sun. Once it was discovered, why did scientists believe that other elements with similar chemical properties existed?

3. Let's go on a scavenger hunt through the periodic table. How many of the following can you find?
   (a) Two representative elements were predicted by Mendeleev but were discovered years later.
   (b) Two elements are named after continents.
   (c) Four elements are named for countries. (Clue: One is an ancient name for one of the other three countries.)
   (d) Four elements are named after a town in Sweden.
   (e) One element is named after a modern city and three after the ancient names of modern cities.
   (f) At least five elements are named after famous scientists. (Bonus points if you find a sixth one!)
   (g) Chromium's name comes from the Greek word *chromos* meaning colour. Find six elements whose names refer to colour or colours. (Bonus points if you find a total of ten!)
   (h) Four elements are named for planets. Since selenium is named after *Selene*, the moon, cerium after *Ceres* and palladium after *Pallas*, both minor planets, they do not count!
   (i) One element is named for a river.
   (j) Five elements are named after mythological gods from Greece, Rome, or Scandinavia. (Bonus marks if you find seven!)
   (k) The name of one element means stench.

P O I N T S · T O · R E C A L L

The atomic mass of an element is the average mass of an atom of the element. The atomic mass of an element appears in the box for that element in the periodic table.

The elements are ordered according to their atomic numbers on the periodic table.

Atoms of the same element have the same number of protons but can have different numbers of neutrons. Atoms of the same element with different numbers of neutrons are called isotopes. Isotopes are usually identified by their symbol followed by their mass number.

The mass number of an isotope is the total number of protons and neutrons found in the nucleus of one atom of that isotope.

Isotopes of an element behave the same way in chemical reactions.

Number of neutrons in the nucleus of one atom of an isotope = mass number − atomic number.

Every element has several isotopes. Some occur naturally while others are formed artificially.

The atomic mass of an element can be calculated from the mass number and percent abundance of each naturally occurring isotope.

Radioactivity is the process in which an unstable nucleus splits apart producing nuclei of other elements, particles and rays, and a lot of energy. Gamma rays are more penetrating than beta particles, and beta particles are more penetrating than alpha particles.

Everything is exposed to background radiation. Three principles to minimize radiation effects are shielding, distance, and time.

Nuclear power plants use a nuclear fuel in a reactor to produce the heat that turns water into steam. The steam runs electric generators. Essential parts of a nuclear reactor are nuclear fuel, moderator, coolant, and control rods.

Humans or animals exposed to a high dose of radiation suffer from radiation sickness.

- Nuclear weapons and possible dangers of nuclear power plants are controversial issues.
- Radioactive isotopes have many industrial, scientific, and medical uses.
- The periodic table has horizontal rows called periods and vertical columns called groups.
- Elements within a group are a chemical family and have similar chemical properties.
- The elements in Group 1 are called the alkali metals, those in Group 2 the alkaline earth metals, those in Group 7 the halogens, and those in Group 8 the noble gases.
- Xenon and krypton are two noble gases known to form compounds with other elements.
- Within a group, the atoms of the elements get larger as we move down the group.
- The periodic table divides elements into metals and nonmetals. Metalloids are found along the jagged line that divides metals and nonmetals.
- The metals generally become more reactive as we move down a particular group and as we move from right to left within the periodic table.
- The nonmetals generally become more reactive as we move up a particular group and as we move from left to right within the periodic table.
- The periodic table can be used to predict chemical and physical properties of elements that have not yet been discovered. These properties should always be confirmed with an experiment.
- Elements in the same group have the same number of electrons in their outermost occupied energy level.
- The lack of reactivity of noble gases has something to do with the presence of an octet of electrons in their outermost occupied energy level. The only exception is helium, which has the maximum of two electrons in its outermost occupied energy level.
- Elements in the same period have the same number of occupied electron energy levels in their atoms.

R E V I E W · Q U E S T I O N S

1. Without referring to the periodic table, give the symbol used for each of the following elements.
   - (a) helium
   - (b) beryllium
   - (c) boron
   - (d) chlorine
   - (e) iodine
   - (f) calcium
   - (g) potassium
   - (h) iron
   - (i) lead
   - (j) gold

2. Without referring to the periodic table, give the name of each of the following elements.
   - (a) Hg
   - (b) Na
   - (c) Br
   - (d) Mg
   - (e) Si
   - (f) P
   - (g) S
   - (h) Cu
   - (i) Sn
   - (j) Ag

3. How can the periodic table be used to indicate which elements are metals and which are nonmetals?

4. Where are metalloids located in the periodic table?

5. List the first 20 elements in the periodic table under the headings *solid*, *liquid*, and *gas*. Write down the symbol and the colour of each element.

6. Copy the following table in your notebook and complete it.

Isotope	Number of protons	Number of electrons	Number of neutrons
hydrogen -1			0
	19		20
uranium -238			
	82		126
		13	14

7. In what way(s) are isotopes of an element
   - (a) the same?
   - (b) different?

8. Identify two isotopes of the same element.

9. Do different isotopes of an element behave in the same way in chemical reactions? Explain.

10. Figure 36 is a box from a typical periodic table.

Fig. 36

19	39.1
K	
Potassium	

   - (a) List three pieces of information you can obtain from this box.
   - (b) Can you tell how many naturally occurring isotopes this element has?

11. What happens when a radioactive nucleus decays?

12. (a) List three types of radiation.
    (b) Which type is the least penetrating?
    (c) Which type is the most penetrating?

13. What is background radiation? List four sources of background radiation.

14. Describe three precautions to take in order to minimize the effects of radiation if you have to work with radioactive isotopes.

15. What are the main parts of a nuclear reactor? Research and describe the role of each of these parts.

16. List at least five beneficial uses of radioactive isotopes.

17. Draw Bohr-Rutherford diagrams to show the electron arrangement of the following atoms:
    (a) neon-20   (b) argon-40
    List two similarities you notice about these two diagrams.

18. What happens to the sizes of atoms as you go down a vertical column on the periodic table? Explain why this is so.

19. Suppose a new element, with atomic number 118, was discovered or made.
    - (a) In which group of elements in the periodic table would this new element belong?

(b) Predict two properties of this element.

20. Which of the following elements are metals, which are nonmetals, and which are metalloids?

titanium (Ti)   hydrogen (H)
lead (Pb)   iodine (I)
aluminum (Al)   sulfur (S)
silicon (Si)   uranium (U)
mercury (Hg)

21. Which elements are in the same family of elements as arsenic?

22. Using only the general trends for reactivity in the periodic table, select the less reactive element from each pair of elements below:

C or N   Fe or Cu   Ca or Zn
Ba or Mg   Ne or F   Au or Cu
Cs or Sr   Cl or F
S or Si   Se or Te

23. The answers you gave in the question 22 are only predictions. As a scientist, what would you have to do in order to check your predictions?

24. (a) To which row of the periodic table does the lanthanide series of elements belong?

(b) Why is the lanthanide series not drawn as part of this row?

5 MEET THE MOLECULE

CONTENTS

| Investigation | 5.1 The Nature of a Molecule |

Would you be able to identify the element copper if you saw it (Figure 1)? Chances are you could because it has a distinctive colour. What tests could you perform to identify copper? The gaseous element oxygen also has characteristic properties that help to identify it. Can you think of one test which you could use to identify oxygen gas?

Fig. 1 *Copper roofs change colour as they tarnish.*

In an earlier chapter, you combined the elements copper and oxygen to form a compound we called copper oxide. How did we know that a compound had formed? Did the compound resemble in any way copper or oxygen? Why not?

Both copper and oxygen are elements, and the smallest unit of an element is an atom. The smallest unit of a compound such as copper oxide is a molecule. If elements combine to form compounds, then atoms combine to form molecules. What is the model for a molecule of copper oxide? How do you show that it is a pure substance and not a mixture of copper atoms and oxygen atoms? How many copper atoms will you need? How many oxygen atoms? These are all good questions.

Draw a model of a copper oxide molecule in your notebook. We will now look at molecules more closely to find out if your model meets all the criteria for a good model. If your model does not, of course, you can always modify it. Developing better models is a fundamental part of science.

Let's begin to study molecules by carrying out the **electrolysis** of a very common substance, water. Electrolysis is the process of decomposing a compound by means of electricity.

The suffix -*lysis* always refers to a decomposition or a disintegration. This suffix comes from the Greek and means a loosening.

Materials

safety goggles
two 10-mL graduated tubes
600-mL beaker
ring stand
2 clamps
stirring rod
2 stoppers (for test tubes)
2 electrodes
9-V storage battery
wooden splints
distilled or deionized water
dilute sulfuric acid

Method

1. In a beaker, add 6 mL of dilute sulfuric acid to 450 mL of the distilled water and stir the solution well.

2. Fill each of the graduated tubes to the brim with this solution. Leave the remaining solution in the beaker.

3. With your forefinger over the opening of a tube, carefully invert the tube into the beaker.

4. Clamp the tube into place using the ring stand and a clamp.

5. Repeat steps 3 and 4 for the second graduated tube (Figure 2).

6. If you inverted the tubes carefully, there should be no bubbles of air in the top of the inverted tubes. If there are air bubbles, try again.

Fig. 2 *The electrolysis of water.*

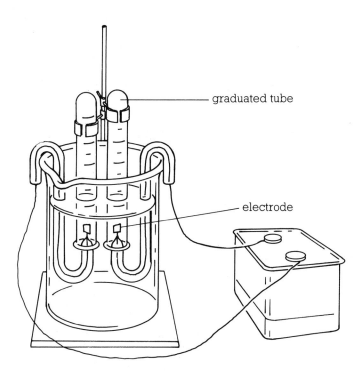

graduated tube

electrode

7. Insert an electrode at the mouth of each tube.

8. Connect one electrode to the positive terminal and the other to the negative terminal of your battery.

Remember that the positive electrode is called the *anode* and the negative electrode the *cathode*.

Follow-up

1. What happens in each tube?

2. What is the volume of gas produced in each tube?

3. Why is there more gas in one of the tubes than in the other?

4. One electrode is connected to the negative terminal; the other, to the positive terminal.
   (a) Which electrode is in the tube with more gas?
   (b) Which electrode is in the tube with less gas?

5. After disconnecting the electrodes, carefully lift the tubes out of the solution. Keep them inverted to trap the lighter-than-air gases inside. Stopper each tube and then turn it upright. Remove the stopper and quickly bring a lighted splint to the mouth of each tube. Record what happens. What gases were produced in this electrolysis?

6. Of the two gases you identified in question 5, which one was formed in greater volume by the electrolysis of water?

7. Why is there a different gas produced at each electrode? Discuss your answer with your classmates.

8. From your observations, what sort of a process or change is electrolysis? Explain your answer.

9. Draw a model of a water molecule. Use the appropriate symbol for each element.

5.2 Applying the Octet Rule

Before we go any further, let's jog our memories a bit. How much can you remember about the elements from Chapter 4?

• Which family of elements is known for its unwillingness to react?

Fig. 3 *This spider has an octet of legs. How many legs are there?*

- What is the name for this family of elements?
- Where is this family located in the periodic table?
- What is the number of the group in the periodic table?
- What does this number tell us about the number of electrons in the outermost occupied shell of the atoms in that group? Is there an exception among the elements in this group? Why is this element an exception? In what way is it like the other elements in the group "in spirit" if not "to the letter"?
- What rule is based on the special stability of these elements (Figure 3)?

Now we are ready to return to our study of the water molecule.

As you learned in investigation 5.1, water is decomposed into the elements hydrogen and oxygen by electricity. Locate the element hydrogen in the periodic table. To which group does it belong? How many electrons are in the outermost occupied, in fact the only occupied, shell or energy level of a hydrogen atom? How many electrons does this atom need to achieve the stability of the nearest noble gas? Think about hydrogen atoms bonding to each other. If a hydrogen atom were to pair up its electron with that of another hydrogen atom, a shared pair of electrons or a chemical bond would be created. How many bonds would each hydrogen atom have to form in order that its outermost occupied shell resemble that of the nearest noble gas? Compare your answer and explanation with those of your classmates.

Now locate oxygen in the periodic table. To which group does oxygen belong? How many electrons are in the outermost occupied shell of the oxygen atom? How many electrons are needed to achieve the stability of the nearest noble gas? How many bonds can an oxygen atom form? Again, discuss your answer and explanation.

Scientists believe that an oxygen atom can form two bonds and a hydrogen atom can form one bond. Therefore, it seems reasonable to predict that, in water, an oxygen atom will form one bond with each of two hydrogen atoms. Is this what you predicted in your model of the water molecule in question 9 of section 5.1? How does the bonding ability of the two elements relate to the volumes of gases produced when water decomposes? Discuss your answer in class.

A link between two atoms in a molecule is called a **chemical bond**. The simplest type of bond involves the sharing of a pair of electrons between two atoms.

Atoms of the elements in Groups 4 through 7 are halfway or more to their goal of achieving an octet of valence electrons. The number of electrons they need to form the octet and, consequently, the number of bonds they can form, is given by the simple formula:

Number of bonds = 8 − group number

H_2O

Questions

1. From their position in the periodic table, determine the number of bonds that the following elements can form:

chlorine	fluorine	nitrogen	carbon
sulfur	phosphorus	silicon	bromine

2. Draw a model for a molecule of the compound formed between hydrogen and *each* of the elements listed in question 1.

3. How did you apply the octet rule to draw models of the compounds in question 2?

4. Fluorine, chlorine, and bromine are all in Group 7 of the periodic table. Each forms a molecule with hydrogen.
   (a) Compare your models of these three molecules. How are the models similar?
   (b) On the basis of the three models in (a), draw a model of the molecule of a hydrogen and iodine compound.

5. For elements in some main groups, the number of bonds is equal to the group number. For elements in other main groups, the number of bonds is equal to 8 minus the group number. For which main group does either equation predict the number of bonds?

5.3 Metals and the Octet Rule

Fig. 4 *Aluminum is a metal with many uses.*

So far, we have neglected to discuss around 70% of all the elements —the metals! Metals also form compounds.

Where are metals located in the periodic table? Which groups in the periodic table consist solely of metals? What information about the valence shell does the group number of metals indicate?

Nonmetals generally have more valence electrons than metals and only need a few electrons more to complete their octet. Most main group metals have less than four valence electrons. The number of bonds that a metal atom with less than four valence electrons can form is equal to the number of the metal's group.

From their location in the periodic table, determine how many bonds the following elements can form:

sodium   calcium   aluminum   magnesium   potassium

The number of bonds that an atom of an element—metal or nonmetal—can form is called the **valence** of the element.

The valences of elements are useful in predicting the formulas of compounds. For example, we know that the valence of sodium is 1 and the valence of chlorine is 1. That means that each sodium atom can form one bond and each chlorine atom can form one bond. When one sodium atom and one chlorine atom bond, then the octet rule is satisfied. The molecular formula of the compound containing sodium and chlorine is NaCl.

Magnesium, on the other hand, has a valence of 2. Each magnesium atom can form two bonds. If magnesium forms a compound with chlorine, which has a valence of 1, then the formula for a molecule of the compound is $MgCl_2$.

What would be the molecular formula for the compound formed from the elements aluminum and chlorine? sodium and oxygen? sodium and nitrogen? aluminum and nitrogen? Keep in mind that, in each case, chemists write the symbol of the metal first, as in $MgCl_2$.

The shell that holds the valence electrons is called the **valence shell**.

A **molecular formula** of a compound shows the number of atoms of each element in a molecule of the compound.

The molecule $MgCl_2$ contains one atom of magnesium and two atoms of chlorine. The subscript "2" in the formula indicates the number of chlorine atoms. The fact that the symbol for magnesium has no number after it means that there is one magnesium atom.

Questions

1. Each of the following pairs of metal and nonmetal elements can be combined to form a compound. Predict the molecular formula for each compound. (Remember to use the position of the element in the periodic table to determine the valence or number of bonds which the element can form.)
   (a) lithium and nitrogen
   (b) sodium and sulfur
   (c) calcium and chlorine
   (d) strontium and sulfur
   (e) barium and fluorine
   (f) aluminum and oxygen
   (g) potassium and iodine
   (h) magnesium and nitrogen

2. Each of the following pairs of elements can be combined to form a compound. Predict the molecular formula for each compound.
   (a) potassium and phosphorus
   (b) lithium and bromine
   (c) carbon and chlorine
   (d) silicon and fluorine
   (e) aluminum and fluorine
   (f) nitrogen and iodine
   (g) carbon and oxygen
   (h) nitrogen and oxygen

3. Can the octet rule be used to explain the following chemical formulas? Give a reason for each answer.
   (a) CO   (c) FeO   (e) BF_3
   (b) NO   (d) Fe_2O_3 (f) BrF_3

5.4 Formulas and Names of Binary Compounds

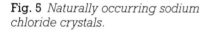

Fig. 5 *Naturally occurring sodium chloride crystals.*

Sections 5.2 and 5.3 introduced the idea of a molecular formula. This formula represents the number of atoms of each element in one molecule of a substance. For example, the compound with the molecular formula NCl_3 has one nitrogen atom and three chlorine atoms in each molecule.

If you completed sections 5.2 and 5.3, you were able to predict formulas by means of the octet rule and drawings of models. But scientists don't want to work through these steps every time they write a formula. Are there quicker methods that we can use?

Consider first the simplest compounds, containing only two elements. They are known as **binary** compounds. Examples include carbon dioxide (CO_2) and sodium chloride (NaCl) (Figure 5). By international agreement, the chemical names of binary compounds end with the suffix *-ide*.

The names of certain very common binary compounds such as water (H_2O) and ammonia (NH_3) are exceptions to this rule.

1. *Compound of two nonmetals*

When a compound is formed from two nonmetals, the name of the compound spells out the formula for you. The prefix that precedes the name of an element tells you the number of atoms of that element in a molecule of the compound. The following table lists the most common prefixes used in naming chemicals.

Prefix	Meaning	Example
mono-	one	carbon monoxide, CO
di-	two	carbon dioxide, CO_2
tri-	three	sulfur trioxide, SO_3
tetra-	four	carbon tetrachloride, CCl_4
penta-	five	diphosphorus pentoxide, P_2O_5
hexa-	six	sulfur hexafluoride, SF_6

The prefix *mono-* is omitted before the name of the first nonmetal. Also, we sometimes shorten a prefix if the result sounds better. For example, "monoxide" sounds better than "mono-oxide."

When you are given the name of a compound containing two nonmetals, look for prefixes that will tell you the number of atoms of each element in a molecule. Several examples are given in the following table.

Name of compound	Number of atoms	Formula
dinitrogen **tri**oxide	2 N 3 O	N_2O_3
sulfur **tetra**chloride	1 S 4 Cl	SCl_4
phosphorus **penta**fluoride	1 P 5 F	PF_5
dinitrogen **mono**xide	2 N 1 O	N_2O

2. *Compound of a metal and nonmetal*

The names for this type of compound are less helpful in providing formulas. The formula of sodium chloride, NaCl, seems fairly easy to predict, although you might wonder why this compound is not called "sodium monochloride." Magnesium chloride, however, has the formula $MgCl_2$. The name gives us no indication that there are two chlorine atoms present for each magnesium atom. In fact, for a binary compound of a metal and a nonmetal, *prefixes are not used in the name*.

How then can we predict the formula of a compound like magnesium chloride? Perhaps the valence of each element in the compound will help. Remember that the valence of an element is the number of bonds an atom of that element can form. The valence of an element in Group 1, 2, or 3 equals the group number. The valence of an element in Group 4, 5, 6, or 7 equals 8 minus the group number. Use the periodic table to determine the valences of magnesium and chlorine. Now look at the formula of magnesium chloride, $MgCl_2$, and see how the formula is related to the valences. Suggest a way of predicting formulas from valences and discuss your suggestion with your group.

Test your suggestion by checking that you can correctly predict the formulas of aluminum bromide ($AlBr_3$) and calcium phosphide (Ca_3P_2). Modify your suggestion as necessary. Now try calcium sulfide (CaS). Do you need to modify your suggestion in any way? Try aluminum phosphide (AlP). Does your modified suggestion work?

You have discovered a method that scientists refer to as the **crossover rule**. We can summarize this rule in six steps.

Metal: Nonmetal:	lithium fluorine		sodium oxygen		magnesium oxygen		aluminum sulfur	
Step 1 Write down the symbols of the elements (metal first).	Li F		Na O		Mg O		Al S	
Step 2 Record the valence value for each element given.	1 Li	1 F	1 Na	2 O	2 Mg	2 O	3 Al	2 S
Step 3 Cross over the valences.	Li_1 F_1		Na_2 O_1		Mg_2 O_2		Al_2 S_3	
Step 4 Find the highest factor common to the two valences.	1		1		2		1	
Step 5 Divide the two valence values by the highest factor.	Li_1 F_1		Na_2 O_1		Mg_1 O_1		Al_2 S_3	
Step 6 Drop any "1" in the formula and write the remaining numbers as subscripts.	LiF		Na_2O		MgO		Al_2S_3	

Writing the chemical name given a formula is quite easy for a compound containing a metal and a nonmetal. Simply write the name of the metal first and add the suffix *-ide* to the name of the nonmetal. DO NOT USE ANY PREFIXES! The following chart gives some examples.

It is not clear from the periodic table, but hydrogen is a nonmetal.

Formula	Chemical name
Na_2S	sodium sulfide
$AlCl_3$	aluminum chloride
Ca_3N_2	calcium nitride
BaH_2	barium hydride

Questions

1. The following compounds contain two nonmetals. Write the chemical formula for each compound.
   (a) sulfur dichloride
   (b) dinitrogen trioxide
   (c) dichlorine monoxide
   (d) boron trifluoride
   (e) silicon tetrachloride
   (f) dinitrogen pentoxide

2. The following compounds contain a metal and a nonmetal. Write the chemical formula for each compound.
   (a) magnesium oxide
   (b) sodium nitride
   (c) potassium sulfide
   (d) aluminum iodide
   (e) strontium fluoride
   (f) barium phosphide

3. The following compounds contain two nonmetals. Give the chemical name for each compound.
   (a) CS_2 (c) SiH_4 (e) N_2O_3 (g) Si_2F_6
   (b) PCl_3 (d) BrF_5 (f) ICl

4. The following compounds contain a metal and a nonmetal. Give the chemical name for each compound.
   (a) Li_2S (c) Al_2S_3 (e) K_3P (g) SrH_2
   (b) CaO (d) $BaBr_2$ (f) NaF

5. Consult the labels on several commonly used foods, pharmaceutical items, or industrial products.
   (a) List the names of ten ingredients found in one or more of those consumer goods.
   (b) Write the chemical formula for each of these ingredients. (If the naming rules you have used so far are not sufficient, it may be necessary to research the chemical formula for some of the ingredients on your list.)

6. Identify the chemical name for the following items usually known by their brand names. You may need to do some research.

   (a) muriatic acid
   (b) lime
   (c) silica gel
   (d) battery acid
   (e) dry ice
   (f) TNT
   (g) rubbing alcohol
   (h) alum
   (i) milk of magnesia
   (j) baking soda
   (k) lye
   (l) photographic "fixer"
   (m) ASA
   (n) washing soda
   (o) carborundum
   (p) rock salt
   (q) MSG
   (r) methylated spirits

7. Write the chemical formula of each substance in question 6. (Use research sources where necessary.)

8. You can now predict the formula of a binary compound of a metal and a nonmetal or of two nonmetals. We have not discussed compounds of two metals. Why not?

5.5 Lewis Diagrams

From the standpoint of the chemist, what do you think is the most important part of the atom? Is it the nucleus?

The composition of the nucleus of an atom identifies the element and that is important. However, it does not tell you whether that atom is part of an element or part of a compound. This distinction is extremely important for the chemist. After all, atoms lose their elemental properties when they become part of a compound. Think of the difference between the properties of water and those of the two elements that make it up, hydrogen and oxygen.

We are looking then for something more than the nucleus. Perhaps the arrangement of electrons around the nucleus? You are getting warmer! Which shell is most responsible for the physical and chemical properties of an element? Which shell's electrons

Remember that electrons occupy energy levels, or shells, around the nucleus.

determine that element's position in the periodic table? Yes — it is the outermost occupied or valence shell of electrons that has the greatest effect on the behaviour of atoms.

In order to focus on this most important part of the atom, Gilbert Lewis suggested that we simply represent an atom with a diagram that includes its symbol and a set of dots around the symbol. These dots indicate only the valence electrons of the atom, since it is these electrons that control the atom's chemical properties. This diagram is known as a **Lewis diagram**, and it is much easier to use than the Bohr-Rutherford diagram. Here are some examples of Lewis diagrams:

$$\text{Li} \cdot \quad \cdot \text{Be} \cdot \quad \cdot \dot{\text{B}} \cdot \quad \cdot \dot{\text{C}} \cdot \quad \cdot \dot{\text{N}} \cdot \quad \ddot{\text{:O}} \cdot \quad \ddot{\text{:F:}} \quad \ddot{\text{:Ne:}}$$

Fig. 6 *In 1916, Gilbert Lewis devised a simple way of representing the valence electrons of atoms.*

Note that if there are four valence electrons or less, the dots are drawn in up to four different locations around the symbol. We say that the individual electrons in each location are **unpaired**. If there are more than four valence electrons, we must begin to draw **electron pairs**. However, we always keep as many electrons unpaired as possible. For a noble gas, other than helium, the Lewis diagram shows four pairs of electrons around the symbol for the atom.

All four locations for the dots are equivalent, so that you may choose any location. For example, the two valence electrons in beryllium, Be, can be drawn in adjacent locations or opposite each other in several ways. Four of the possibilities are:

$$\overset{\bullet}{\text{Be}} \cdot \qquad \overset{\bullet}{\underset{\bullet}{\text{Be}}} \qquad \cdot \text{Be} \cdot \qquad \underset{\bullet}{\text{Be}} \cdot$$

Similarly, the two pairs in the diagram for oxygen can be adjacent to or opposite each other. For example:

$$\cdot \overset{\bullet\bullet}{\underset{\bullet\bullet}{\text{O}}} \cdot \qquad \cdot \overset{\bullet}{\underset{\bullet\bullet}{\text{O}}} \text{:}$$

Some elements do not exist in nature as lone atoms. They always appear in pairs like mittens or shoelaces. These are the so-called **diatomic** elements. A useful mnemonic for remembering these elements is:

"Diatomic" means having two atoms per molecule.

I Have No Bright Or Clever Friends

The first letters of the words in this sentence are the symbols for the elements whose atoms pair up in diatomic molecules.

$$I_2 \quad H_2 \quad N_2 \quad Br_2 \quad O_2 \quad Cl_2 \quad F_2$$

Locate these elements on your periodic table. To which category of elements do they belong? In which "neighbourhood" of the periodic table do they "live"? As in compound formation, atoms of these diatomic elements try to get an octet of electrons in their valence shell. There is one exception—can you guess which element it is? What number of valence electrons will make one atom of that element stable? In other words, what is that element's goal?

To complete their octets, atoms with unpaired electrons link up to form a **bonding pair** of electrons, which the atoms share. For example, a fluorine atom has seven electrons in its valence shell. Therefore, it has one unpaired electron ready for bonding with another fluorine atom.

$$:\ddot{F}\cdot \quad + \quad \cdot\ddot{F}: \quad \longrightarrow \quad :\ddot{F}:\ddot{F}:$$

Note that in the molecule held together by the bonding pair, each fluorine atom has eight valence electrons around it. Before the two atoms combined, each one had only seven valence electrons.

Draw the Lewis diagrams for the diatomic molecules of the other halogens. How should the hydrogen molecule be drawn?

An atom with more than one unpaired electron can form more than one bond. An example of this is an oxygen atom, which has two unpaired electrons in its valence shell.

As with most rules, the ones governing Lewis diagrams have some exceptions. Experimental evidence suggests that oxygen molecules retain two unpaired electrons. The following Lewis diagram reflects this observation:

$$:\ddot{O}:\ddot{O}:$$

$$:\ddot{O}\cdot \quad + \quad \cdot\ddot{O}: \quad \longrightarrow \quad :\ddot{O}::\ddot{O}:$$

Draw the Lewis diagram for the diatomic nitrogen molecule.

The elements phosphorus and sulfur are **polyatomic**. This means that molecules of these elements contain several atoms. Phosphorus molecules can be represented by the formula P_4 and sulfur molecules by the formula S_8.

In keeping with the valence of phosphorus, each atom of phosphorus in the molecule is bonded to three other phosphorus atoms. This can be achieved if the molecule is shaped like a three-sided pyramid with a phosphorus atom at each apex.

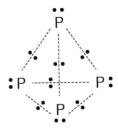

How many bonds can a sulfur atom form? How would that affect the shape of the sulfur molecule? Draw the Lewis diagram for the S_8 molecule in your notebook.

Lewis diagrams can be used to represent molecules of compounds, not just molecules of elements. You can draw the Lewis diagram of water (H_2O), remembering that an oxygen atom has two unpaired electrons ready for bonding and that each hydrogen atom needs one more electron to complete its valence shell.

$$H\cdot \ + \ \cdot\ddot{\underset{..}{O}}\cdot \ + \ \cdot H \ \longrightarrow \ H\!:\!\ddot{\underset{..}{O}}\!:\!H$$

Because the four locations around the oxygen atom are all equivalent, it does not matter whether the two hydrogen atoms are drawn adjacent to or opposite each other.

Questions

1. Draw Lewis diagrams to represent molecules of the following binary compounds.
   (a) hydrogen chloride (HCl)
   (b) ammonia (NH_3)
   (c) methane (CH_4)
   (d) sodium chloride (NaCl)
   (e) potassium iodide (KI)
   (f) magnesium oxide (MgO)

2. What advantages does the Lewis diagram have as a way of representing the structure of matter?

3. Does the fact that you can draw an acceptable Lewis diagram for a molecule prove that atoms are held together in the way the diagram suggests? Explain.

4. Why should the sharing of electrons between atoms hold the atoms together? (*Hint:* Think about electrostatic charges.)

5.6 Structural Formulas

To further simplify the Lewis diagram, each electron pair that is *between* the atoms is replaced by a single line. When we do this, we change a Lewis diagram into a **structural formula**. For example, consider the structural formula of the hydrogen molecule:

$$H\!:\!H \ \ \dashrightarrow \ \ H\!-\!H$$

Similarly, the Lewis diagram for diatomic fluorine becomes:

$$:\ddot{F}:\ddot{F}: \quad \dashrightarrow \quad :\ddot{F} - \ddot{F}:$$

And for the oxygen molecule the Lewis diagram becomes the structural formula:

$$:\ddot{O}::\ddot{O}: \quad \dashrightarrow \quad :\ddot{O} = \ddot{O}:$$

The Lewis diagrams for binary compounds can also be simplified into structural formulas.

$$H:\ddot{O}:H \quad \dashrightarrow \quad H - \ddot{O} - H$$

Questions

1. For each of the following pure substances:
   (a) write the molecular formula, and
   (b) draw the structural formula.
      (i) diatomic nitrogen   (vi) ammonia
      (ii) diatomic chlorine   (vii) methane
      (iii) diatomic bromine   (viii) sodium chloride
      (iv) diatomic iodine   (ix) potassium iodide
      (v) hydrogen chloride   (x) magnesium oxide

2. For the following pure substances:
   (a) write the molecular formula, and
   (b) draw the structural formula.
      (i) nitrogen trifluoride
      (ii) carbon dioxide
      (iii) silicon tetrachloride
      (iv) dichlorine monoxide
      (v) hydrazine (which has two N atoms
         and four H atoms per molecule)
      (vi) phosphorus tribromide
      (vii) sulfur dichloride

3. Compare the advantages and disadvantages of the Lewis diagram and the structural formula as representations of matter.

4. Does the fact that you can drawn an acceptable structural formula for a molecule prove that atoms are held together in the way the diagram suggests? Explain.

5. Use your imagination and devise another way to represent the structure of pure substances.

5.7 Making Three-Dimensional Models of Pure Substances

Notice the connecting "sticks," which serve as the bonds connecting atoms in your three-dimensional structures. Pay attention to the colour coding used to identify models of different kinds of atoms. These model atoms take into account the common shapes of molecules.

Using a molecular model kit, build three-dimensional models of the molecules of hydrogen, nitrogen, oxygen, phosphorus, and sulfur.

How do these three-dimensional models compare to the structural formulas of molecules for these elements?

Build three-dimensional models of the following binary compounds: hydrogen chloride, water, ammonia, methane, sodium chloride, potassium iodide, and magnesium oxide. Compare these three-dimensional models to the structural formulas you drew for question 1 of section 5.6. What important differences do you notice?

Questions

1. Using your molecular model kits, build three-dimensional representations of the following binary compounds.
   (a) nitrogen trifluoride
   (b) carbon dioxide
   (c) silicon tetrachloride
   (d) dichlorine monoxide
   (e) hydrazine (see section 5.6, question 2)
   (f) phosphorus tribromide
   (g) sulfur dichloride

2. What are the advantages and disadvantages of using molecular model kits to represent the structures of pure substances?

3. (a) In what way are these molecular models an improvement over Lewis diagrams?
   (b) In what way are they less convenient than Lewis diagrams?

4. Does the fact that you can build a molecular model of a molecule prove that the atoms are held together in the way the model suggests? Explain.

5. Consider the compounds formed between hydrogen and each of the elements in row 3 of the periodic table, beginning with sodium. For each compound:
   (a) Predict the formula.
   (b) Draw the Lewis diagram.
   (c) Draw the structural formula.
   (d) Build the molecular model.

6. Examine the Lewis diagram and structural formula for water on page 134. Build a molecular model of a water molecule and compare its shape with the shape suggested by the Lewis diagram and structural formula. Can you tell which shape is correct? Explain.

5.8 Representing Chemical Reactions

You just received a letter from John Dalton! He wants to know more about the electrolysis experiment you performed recently. He has been under the impression that the formula for water was HO and your experiment seems to contradict his idea. He wants you to describe what happened in the electrolysis. It is important that you mention the reactants, the products, and any special conditions associated with the experiment. Get together in small groups and discuss how you will respond to his question. Prepare a paragraph describing the electrolysis reaction and share it with your classmates.

Most likely your paragraph was several sentences long. You probably wished there was some way to say the same thing more briefly. There is — it is called the **chemical equation**. In a chemical equation, the formulas of the reactants are written, separated by plus signs, on the left of the reaction arrow. The formulas of the products, also separated by plus signs, appear on the right.

In an earlier chapter of this module, you learned to write word equations. In your notebook, write the word equation that describes the electrolysis of water. Write on top of the arrow any special conditions required for the reaction to occur. (In this case, you could write "electricity.")

Replace each of the names in the word equation with the appropriate chemical formula.

Using a molecular model kit, represent this reaction in three dimensions. Do you run into any difficulties with the equation as it is written?

How would you indicate the observation that twice as much hydrogen gas was produced as oxygen gas?

Does the equation make sense now? (Check the number of atoms of each element shown in the equation.) If the equation does not make sense, how can you modify it so that it does?

When you put the lighted splint into the tube containing hydrogen gas, there was a small explosion. This explosion accompanied the formation of water from the hydrogen gas in the tube and the oxygen gas in the surrounding atmosphere. Write a

Recall that a reaction in which two elements or two smaller compounds combine to form a single product is called a synthesis reaction. It is the opposite of a decomposition reaction.

chemical equation for this reaction and include any special conditions.

You probably have noticed that the reaction between hydrogen gas and oxygen gas to form water is the reverse of the electrolysis reaction. How would you summarize both reactions in a single equation?

How would you illustrate the synthesis of water using Lewis diagrams?

Questions

1. Write a chemical equation for the synthesis of each of the following compounds from its elements:
   (a) carbon dioxide
   (b) sodium chloride
   (c) magnesium oxide
   (d) potassium iodide
   (e) ammonia

2. Write a chemical equation for the decomposition of each of the following binary compounds into its elements:
   (a) carbon dioxide
   (b) sodium chloride
   (c) magnesium oxide
   (d) potassium iodide
   (e) ammonia

3. Draw Lewis diagrams to illustrate each reaction in question 2.

4. Build three-dimensional models to illustrate each reaction in question 1.

P O I N T S · T O · R E C A L L

- The molecule is the smallest unit of a compound. A molecule contains two or more atoms that are chemically bonded to one another.
- Electrolysis is the process in which a compound is decomposed by electricity.
- In the electrolysis of water, the compound decomposes into hydrogen gas and oxygen gas.
- The observation that twice as much hydrogen forms as oxygen tells us that the correct formula for water is H_2O.
- The number of bonds that an atom can form is called its valence.
- The valence of elements in the main Groups 1, 2, and 3 is given by their group number in the periodic table. The valence of elements in the main Groups 4, 5, 6, and 7 is equal to 8 minus the group number in the periodic table.
- A binary compound is one that contains only two elements.
- When the name of a binary compound does not indicate its formula, the formula can be determined by crossing over the valences of the elements.
- In a binary compound, the suffix *-ide* is added to the name of the second element.
- To name a binary compound containing two nonmetals, translate the number of atoms of each element into a prefix for the name of that element.
- To name a compound containing a metal and a nonmetal, name the metal first and add the suffix *-ide* to the nonmetal.
- The elements that exist as diatomic molecules are bromine, chlorine, fluorine, oxygen, iodine, hydrogen, and nitrogen.
- Elemental sulfur comes in molecules of eight atoms (S_8) and phosphorus in molecules of 4 atoms (P_4).
- In a Lewis diagram, an atom is represented by the symbol of the element and a set of dots around the symbol to indicate its valence electrons.
- Atoms form bonds to other atoms by linking their unpaired electrons.
- The Lewis diagram of a compound can be simplified by using a line to represent each pair of electrons that is shared between two atoms. The result is called a structural formula.
- A chemical equation is a shorthand way of describing a chemical reaction. The formulas of the reactants appear on the left of the arrow, the formulas of the products on the right. Any special reaction conditions are written above the arrow.
- Chemical equations are balanced by "trial and error." When the equation is balanced, the number of atoms of each element is the same on the left- and right-hand sides of the arrow.

R E V I E W · Q U E S T I O N S

1. What kind of a chemical reaction is an electrolysis?
2. How do we know that the correct formula for water is H_2O?
3. Give the valence of the following elements:

    (a) lithium
    (b) carbon
    (c) iodine
    (d) sulfur
    (e) calcium
    (f) aluminum
    (g) magnesium
    (h) sodium
    (i) chlorine
    (j) phosphorus

4. Predict the formulas of compounds formed from the following pairs of elements:

    (a) carbon and chlorine
    (b) hydrogen and sulfur
    (c) magnesium and phosphorus
    (d) lithium and nitrogen
    (e) aluminum and oxygen

5. Name the following binary compounds formed from two nonmetals:

    (a) SO_3
    (b) OF_2
    (c) N_2O
    (d) CO
    (e) P_2O_5
    (f) SiF_4
    (g) SF_6

6. Name the following binary compounds formed from a metal and a nonmetal:

    (a) NaI
    (b) K_2O
    (c) AlN
    (d) $BeCl_2$
    (e) CaS
    (f) BaF_2
    (g) Li_3P

7. Write the chemical formula for each of the following compounds:

    (a) carbon disulfide
    (b) aluminum bromide
    (c) boron trifluoride
    (d) silicon dioxide
    (e) potassium sulfide
    (f) lithium chloride
    (g) dinitrogen tetraoxide
    (h) magnesium oxide
    (i) diphosphorus trioxide
    (j) strontium iodide

8. What is the common name for each of the following compounds?

    (a) dihydrogen monoxide
    (b) nitrogen trihydride
    (c) carbon tetrahydride

9. Use a Lewis diagram to represent each of the following molecules:

    (a) N_2
    (b) CCl_4
    (c) HBr
    (d) PH_3
    (e) CS_2
    (f) H_2O_2

10. Draw a structural formula to represent each of the following molecules:

    (a) O_2
    (b) H_2S
    (c) NF_3
    (d) CO_2
    (e) N_2H_4
    (f) SiH_4
    (g) ClF
    (h) C_2H_2

11. Represent each of the following synthesis reactions by means of the following:

    (a) chemical equation
    (b) a molecular model kit
    (c) Lewis diagrams

    (i) potassium + chlorine → potassium chloride

    (ii) carbon + oxygen → carbon monoxide

    (iii) hydrogen + iodine → hydrogen iodide

    (iv) nitrogen + hydrogen → ammonia

6 MAGNETISM AND ELECTRO-MAGNETISM

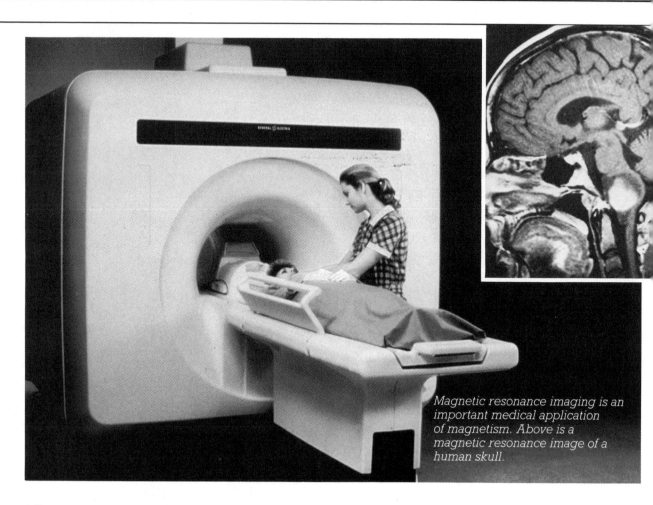

Magnetic resonance imaging is an important medical application of magnetism. Above is a magnetic resonance image of a human skull.

CONTENTS

6.1 Research Project

In section 1.3, you completed a research project on the properties and structure of matter. You can now apply your research skills to another topic—the magnetic and electrical properties of matter. To complete the project, you need some background knowledge about electricity and how it is measured. Your teacher will decide when you have enough knowledge to begin the project. It is probably best not to read the rest of this section until your teacher tells you to.

The project involves the design and construction of a measuring instrument. The aim is then to use the intrument to measure at least one variable associated with an electrical circuit. There are many possible instruments to choose from:

1. Electroscope

2. Galvanometer

3. Ammeter

4. Voltmeter

5. Ohmmeter

6. Multimeter (to measure voltage, current, and resistance)

7. Wattmeter

8. Joulemeter

9. Compass

10. Current balance

11. Thermocouple

12. Electric pyrometer

13. Photometer

14. Some other instrument

For your measurement project to be successful, it must be well organized and have definite goals. Suppose that, instead of working with electricity, you plan to measure mass. One way to do this is to use a beam balance (Figure 1).

Fig. 1 *Beam balances.*

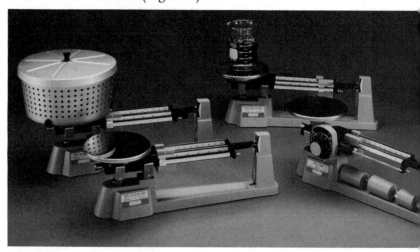

Suppose that you choose to build your own beam balance. How do you go about it, what kinds of information do you need, and what kinds of decisions do you need to make? The following points summarize the main steps in the process:

(a) *Understand scientific measurement.* Scientific instruments are only useful if they give measurements that are reproducible. In other words, measurements of the same quantity should be the same over and over again, to some acceptable level of uncertainty.

Scientists refer to reproducibility as **precision**.

A second requirement of scientific instruments is that they must give measurements that are **accurate**. That is, the measurements must be close to the correct values.

You can learn more about the concepts of precision and accuracy by referring to Appendix 1.

(b) *Choose a variable*. Now that you know the general idea of using scientific instruments, you can decide the goals of this particular project. First, choose the variable(s) you wish to measure. In this project, the variable(s) will be electrical. One possible variable is the electric current. But let us continue to assume that the variable is mass.

(c) *Examine the instrument*. Having chosen the variable, think about the way in which it is measured. Again assume that, having chosen mass as the variable, we decide to build our own beam balance. It then becomes essential to examine the balance closely to see how it works and what it is made of.

(d) *Suggest a procedure*. Having examined the instrument, decide how you would go about building your own. There are many stages to this process. Your teacher may decide to supply you with a checklist to help you.

(e) *Have your procedure approved*. You may need some helpful hints. Also, safety is essential. So do not start to build your instrument until your teacher has approved your procedure.

(f) *Draw a diagram*. When people build things, they often work from sketches. As your instrument is electrical, draw a circuit diagram to represent it. Label the circuit diagram by means of a conventional electrical symbol for each component. Again, have your teacher check your idea for feasibility and safety.

(g) *Build the instrument*. Follow your circuit diagram. List all the materials you need. Be very specific. For example, if the instrument requires a high-resistance wire, is it 10 Ω or 50 Ω? Check that the instrument works. If it does not, modify it and change the circuit diagram.

(h) *Graduate the instrument*. Every measuring instrument needs a scale, which is a series of markings known as graduations. Setting up a scale from scratch can be difficult — you need to choose and name a unit. Suppose that you have built a beam balance, but you have no standard masses to help you make a scale. You do have some coins, so you decide that your standard unit of mass will be the mass of one penny. You call it "the penny" and choose "p" as its symbol. Using one, two, three pennies, and so on, you mark divisions on the scale. As the mass of a penny is quite large for laboratory purposes, you might then decide to subdivide each division on the scale. You might choose to subdivide into tenths and hundredths, representing "decipennies" and "centipennies," with symbols dp and "cp." You could also mark multiples of your basic unit on the scale, if necessary. For example, the mass of ten pennies would be a "decapenny," symbol "dap."

(i) *Test the instrument.* Carry out experiments to see how reliable your instrument is. Compare its precision and accuracy with those of commercially available instruments.

(j) *Improve the instrument.* Suggest and make improvements. Test the performance of the instrument. Modify the circuit diagram for the instrument as necessary.

(k) *Test a scientific law.* In later sections, you will discover various rules and laws that apply to electricity. Test one such rule or law with your instrument. Then, modify the way in which the instrument is graduated, so that the scale shows the conventional unit(s) for the variable(s) you are measuring, instead of the unit(s) you invented in (h). To do this, you will need to find out the relationship between the standard unit(s) and the unit(s) you made up. (For example, suppose you used the mass of a penny as the unit of mass for a beam balance, but you would like to graduate the balance in grams. You need to know the mass of a penny in grams.)

(l) *Write a report.* Include the difficulties you encountered, how you overcame them, and the results you obtained with the instrument you built. Support your research with data tabulated in such a way that they show patterns or relationships. Include any graphs you plotted to obtain mathematical relationships.

| Investigation | 6.2 Magnetism |

We become familiar with magnets and magnetism from an early age. In this course, you have already examined the magnetic properties of substances before and after they underwent changes. (Refer to section 2.2.) In this chapter, you will learn more about magnetism. We can begin with what you already know:

1. From your own experiences, list devices that use or depend on magnetism. Compare your list with those of other students in the class.

2. Answer and discuss the following:
   (a) Do magnets attract all substances?
   (b) Do magnets attract other magnets?
   (c) Do some parts of magnets attract substances more strongly than other parts?

3. How can we classify a selection of materials into two or three categories according to their magnetic properties?

You will now test your answers to questions 2(a) and 3 by means of an experiment. Subsequent activities will allow you to test your answers to 2(b) and 2(c).

Materials

2 bar magnets
paper clips
thumb tacks
needles
paper
small pieces of wood
jewellery
coins
pen
pencil
lodestone

Method

1. Use your own procedure to test the materials.

2. Use a chart to show your classification of the materials.

Follow-up

1. Did any material(s) other than the bar magnets behave like a magnet?

2. Were substances that were attracted by the magnet all attracted to the same degree?

3. Can a bar magnet lift an object without being in direct contact with it? (Design an activity with paper clips and a magnet to answer this question.)

4. From the activity you designed in question 3, will the paper clips attract each other after the magnet is removed?

The ability of a substance to stay magnetic is called **remanence**.

5. Consider the two paper clips that were closest to the magnet in question 3. Separate them and tap both ends of each clip 20 times on the desk. Bring the clips back into contact. Are they still attracted to each other?

6. You answered three questions at the start of this investigation. Do you want to modify any of your answers on the basis of your observations?

7. Physicists describe materials that are strongly attracted to a magnet as **ferromagnetic**. Materials that are not attracted are called **non-magnetic**. Magnets, of course, are "magnetic." Classify the materials you tested into these three groups. Does this classification agree with the one you developed yourself?

8. Discuss your answers to question 7. Are there any disagreements over the category that any material belongs to? Retest any material(s) you are unsure of and try to reach an agreement.

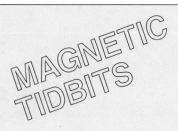

MAGNETS EVERYWHERE

A wide range of magnets is used in the home and in industry (Figure 2). The shapes and sizes of magnets vary with their uses. A very common example in the home is the magnetic strip used on many refrigerator doors. The strip is made of powdered magnetic ceramic embedded in rubber.

Some can openers contain magnets to hold a can lid after cutting it off. Audio and video recorders use magnetic tapes. These are coated with small needles of a suitable material, such as an oxide of iron, cobalt, or chromium. Each needle acts as a small magnet.

How many other uses of magnets can you think of?

Articles made from baked clay are called **ceramic**. Examples are pottery and brick.

Fig. 2 *Video recorders use magnetic tapes.*

Investigation	## 6.3 Magnets and Magnetic Forces

It took many centuries for people to realize that magnets could not only attract some things but could also repel them. In this investigation, you will begin to take a closer look at magnetic forces.

A **force** is a push or a pull on an object.

Materials

2 bar magnets
retort stand
ring clamp
thread

Method

The two ends of a bar magnet are called **poles**.

1. Tie the thread around the middle of one bar magnet and suspend the magnet from the ring clamp (Figure 3a). Ensure that the magnet is not close to another magnet or to a ferromagnetic material. Let the magnet swing freely until it comes to rest. Note the direction in which each pole of the magnet points.

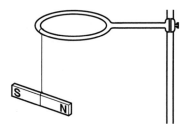

Fig. 3 (a)

2. Hold the second magnet. Slowly bring its north pole (marked N and often called the N-pole) near the N-pole of the suspended magnet (Figure 3b). Record your observation.

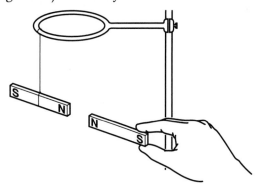

Fig. 3 (b)

3. Repeat step 2, but this time bring the south pole (marked S and often called the S-pole) of the second magnet toward the N-pole of the suspended magnet. Make sure that the magnets are the same distance apart as in step 2. Record your observation.

4. Repeat step 3, but start with the magnets further apart and watch carefully what happens as you bring them closer together. Record your observation.

5. You should already be forming some impressions of the ways in which the two magnets behave. What other tests could you carry out to confirm your impressions? Try them.

Follow-up

1. From your observations, state two qualitative laws of magnetic attraction and repulsion.

2. Do the forces you studied seem to depend on distance? What evidence supports your view?

6.4 Theory of Magnetism

Though you now know some of the properties of magnets, there are many unanswered questions. Some of the more obvious ones are:

What makes one piece of iron a magnet, while another piece may not be?

Why does every magnet have two poles?

Does breaking a magnet into two pieces give two magnets?

Why does tapping a magnet several times demagnetize it?

Can you make a magnet with only one pole?

The answers to all these questions depend on one basic fact: A magnet consists of many sets of particles aligned in the same direction, so that their N-poles point one way and their S-poles the other way (Figure 4a). These sets of particles are called **domains**.

Destroying magnetic properties is called **demagnetizing**.

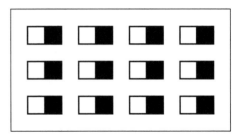

Fig. 4 (a)

In a bar of a ferromagnetic substance, the magnetic domains point in all directions (Figure 4b). The effect of the N-pole of one is offset by the effect of the S-pole of another. The magnetic effects cancel out and there are no "free" poles near the ends of the bar. Aligning the domains in the ferromagnetic bar by some method will make it into a magnet. This is known as **magnetizing** the bar.

Fig. 4 (b)

If a magnet is cut into pieces, each piece is a smaller magnet with N- and S-poles (Figure 4c). However, if you strike a magnet with a hammer, some of the domains lose their alignment and become randomly arranged. So the magnet loses some of its magnetism.

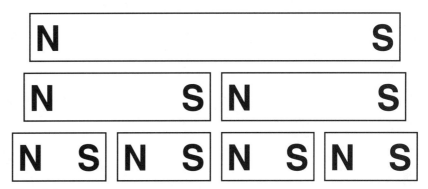

Fig. 4 (c)

You may be wondering about non-magnetic substances. They do not contain magnetic domains, so they are not attracted by magnets and cannot be magnetized.

Questions

1. What happens to the domains in a ferromagnetic substance as a magnet approaches it?

2. From your answer to question 1, why does a magnet attract a ferromagnetic substance?

6.5 Remanence

In investigation 6.2, you noticed that the paper clips remained attracted to each other after the magnet was removed. In the presence of the magnet, some domains within the paper clips must have become aligned to form new magnets. Since the paper clips stayed magnetic after the magnet was removed, the domains could not have become random again. As noted above, a material's ability to stay magnetized is called remanence, also known as **retentivity**.

Ferromagnetic materials differ in how easily their domains align. They align easily in an iron bar. In other words, it is easy to magnetize iron. But, if the domains align easily to form a magnet, they can also move out of alignment easily. Materials like iron, which can be magnetized and demagnetized easily, are referred to as magnetically ''soft.''

MAGNETIC TIDBITS

THE EARTH AS A MAGNET

The ancient Chinese may have been the first to discover that pieces of a type of rock, called lodestone, always came to rest pointing in the same direction when freely suspended. A Chinese emperor, Hwang-Ti, is said to have used a primitive compass on his chariot in about 250 B.C. In 1269, Petros Peregrinus, a French Crusader, gave the first detailed description of a compass similar to those we use today (Figure 5).

But, for a compass to be effective, the magnet inside it must be attracted to something. About 400 years ago, William Gilbert, of Colchester, England, suggested that the Earth itself behaved as if it were a huge magnet. No matter where you are on Earth, something attracts the N-pole of a compass needle one way and the S-pole in the opposite direction.

Any region in which magnetic effects are felt is called a **magnetic field**. The cause of the Earth's magnetic field is still not readily understood, though various theories have been suggested. Its likely cause seems to be electric charges moving within the Earth's liquid core, as a result of the heat produced there by radioactivity. Whatever the reason for the Earth's magnetism, the field produced is similar to the field one might expect if the Earth contained a huge bar magnet.

It is known today that the compass needle points to the Earth's north magnetic pole, which is in northern Canada at a latitude of 73.3°. This is over 2000 km from the geographic North Pole.

Fig. 5 *A simple compass.*

Obviously, iron and other "soft" materials are not the best for making magnets that stay magnetic, known as **permanent magnets**. A better choice is steel, which has a higher retentivity or remanence than iron. In other words, steel is more difficult to magnetize and demagnetize than iron. Steel is an example of a magnetically "hard" material.

Questions

1. Why is steel harder to magnetize and demagnetize than iron?

2. Why did the magnetized paper clips lose their magnetism when they were tapped on the desk?

Investigation | 6.6 *Mapping a Magnetic Field*

A magnetic field is the region around a magnet where magnetic forces of attraction or repulsion are felt. We cannot see a magnetic field with the naked eye, but we can do experiments to show that it is there.

Let's begin with your own hypothesis of what a magnetic field might look like. Draw a diagram showing a bar magnet and how you think the magnetic field around it might look, if you could see it. Compare your drawing with those of others in the class. Then complete the investigation to test your hypothesis.

Materials

bar magnet
magnetized needle pushed through a cork
battery jar
water
iron filing shaker in a plastic bag
safety goggles
sheet of paper

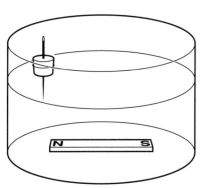

Fig. 6

Method

I. *Mapping with a Compass Needle*

1. Place a bar magnet at the bottom of a jar that is two-thirds full of water.

2. Float the cork on the water, with the marked N-pole of the needle pointing downward near the N-pole of the magnet (Figure 6).

Handle needles with care. Tell your teacher if you prick yourself.

3. Release the cork and observe its motion.

4. Move the cork to the other side of the magnet. Again place it near the N-pole, release the cork, and observe its motion.

(a) Draw a diagram to show the magnet and the direction in which the free N-pole of the needle moved. (Draw lines and put an arrowhead on each one to show the direction of motion.)

(b) The direction of motion indicates the shape of the magnetic field. How does this observed shape compare with the one you drew at the beginning of the investigation?

(c) Physicists refer to the lines of motion that you have drawn as **lines of force**. Every line of force has a shape and a direction. The direction is shown by the arrowheads you drew in (a). So, the direction of the arrow on a line of force is the direction in which a free north pole would move along it. The diagram you drew in (a) does not show much detail. If you were to add more lines of force to the diagram, what do you think they would look like? Draw and discuss your idea. Part II of the method will allow you to test your prediction.

II. *Mapping with Iron Filings*

Do not get iron filings in your eyes. Wear goggles.

1. Place a sheet of paper on top of a bar magnet. Sprinkle iron filings thinly and evenly on the paper.

2. Tap the paper gently with a pencil so that the iron filings show the pattern of the lines of force. Each filing is magnetized and turns in the direction of the field.

3. Draw a sketch of the observed pattern.

4. Mark the direction of each line of force with an arrow. (To find the direction in which a free north pole would move, use a compass needle and observe the direction in which its N-pole points.)

(a) Give two definitions of magnetic lines of force.

(b) State three characteristics of the lines of force around a bar magnet.

(c) You have only examined the force field around a single bar magnet, but you already know that poles of two magnets affect each other when they are brought together. Draw what you think the lines of force would look like in each of cases (i) and (ii). (See Figures 7 and 8.) Then devise an experiment to test your hypotheses.

(i) **Fig. 7**

(ii) **Fig. 8**

| Investigation | ## 6.7 Mapping the Field Around a Straight Conductor |

You have examined the magnetic fields created by permanent bar magnets. But can magnetic fields be created in other ways? What, if anything, happens to a compass needle when it is placed near a straight copper wire that carries an electric current? Is there a magnetic field around the wire? If so, what is its shape in a plane perpendicular to the wire? Assume that a field exists and draw a sketch of what you expect its shape to be. Discuss your sketch with your group. Complete the investigation to test your hypothesis.

Materials

Propose your own materials list when you devise a procedure.

Method

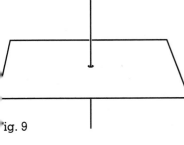

Fig. 9

1. Pass a straight copper wire through a hole in the centre of a piece of cardboard, such that the cardboard is perpendicular to the wire (Figure 9).

2. Have your teacher help you connect the wire to a source of electric current.

3. Develop your own procedure for plotting the magnetic field in the plane perpendicular to the straight, current-bearing wire.

4. Have your teacher check and approve your procedure. Then carry it out.

5. Sketch the shape and direction of magnetic lines of force you observe around the wire.

Follow-up

1. Compare your hypothesis with your observations.

2. If you reverse the direction of the current in the wire, what do you think will happen to each of the following?
   (a) the shape of the magnetic field
   (b) the direction of the lines of force

3. Test your hypotheses from question 2.

4. Let's propose a rule for predicting the direction of the lines of force around a straight conductor. This rule, which we will call the **left-hand rule**, is as follows:

Do not grasp a conductor while an electric current is flowing through it.

Remember that electrons flow out of the negative terminal of a power supply and back to the positive terminal.

"If you grasp the conductor in your left hand, with your thumb pointing in the same direction as the electrons move, then your fingers curl around the conductor in the direction of the lines of force."

Do your findings in the experiment agree with this rule?

Fig. 10

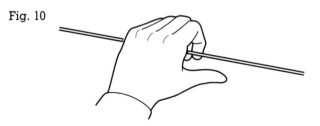

5. Write a report on your findings. Include the answers to the follow-up questions.

Extension

What do you expect the magnetic field to look like in each of the following cases? Test your hypotheses.

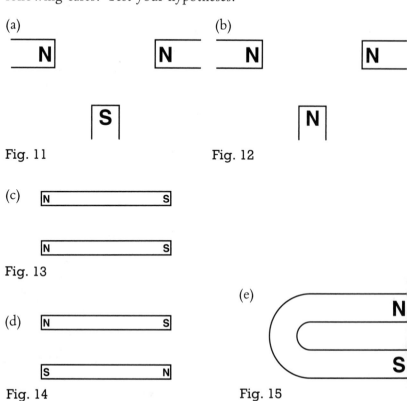

(a)

Fig. 11

(b)

Fig. 12

(c)

Fig. 13

(d)

(e)

Fig. 14

Fig. 15

Investigation

6.8 Mapping the Field Around a Solenoid

Recall that a **solenoid** is a coil of wire wrapped uniformly around a cylinder. An easy way to make a solenoid is to wrap some wire around a small test tube and tape the wire in place.

What kind of magnetic field do you expect a current-bearing solenoid to produce? Sketch and discuss your ideas. Then test your hypothesis.

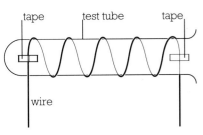

tape   test tube   tape

wire

Fig. 16

Materials

Propose your own materials list when you devise a procedure.

Method

1. Develop your own procedure for plotting the magnetic field around a current-bearing solenoid. Make sure that your procedure will allow you to answer the follow-up questions.

2. Have your teacher check and approve your procedure. Then carry it out.

3. Sketch the shape and direction of magnetic lines of force you observe around the solenoid.

4. Write a report on your findings. Include the answers to the follow-up exercises.

Follow-up

1. Were your predictions about the magnetic field correct? If not, how should you modify your hypothesis?

2. Does the solenoid have magnetic poles? If so, how did you determine where they were and which was which?

3. When you reversed the direction of the current in the wire, what happened to each of the following?
   (a) the shape of the magnetic field
   (b) the direction of the lines of force

4. In your own words, suggest a rule that relates the direction of the electron flow in the solenoid to the direction of the lines of force.

5. Now try to apply the left-hand rule to your solenoid. Turn off the current. Then wrap the fingers of your left hand around the coil in the direction the electrons have been flowing in. The rule

predicts that your thumb points toward what was the N-pole of the solenoid. Is this prediction consistent with the rule you suggested in question 4? Explain.

6. What happened to the magnetic field of the solenoid when the current was turned off?

7. List some devices in your environment that contain solenoids.

| Investigation | 6.9 Strength of an Electromagnet |

Earlier in the chapter, you investigated the field surrounding a permanent bar magnet. You have now seen that a wire carrying an electric current also produces a magnetic field. The magnetism in this case depends on the moving electrons in the wire. We call such magnets **electromagnets**. When the current is turned off, the wires are no longer magnets.

The solenoid you built in the previous investigation consisted of wire wrapped around an empty test tube. What do you think will happen to the strength of the electromagnet if a solid core is placed inside the test tube? Discuss this question with your group and suggest a hypothesis to explain the anticipated change.

Materials

Propose your own materials list when you devise a procedure.

Method

1. Develop your own procedure for measuring the strength of the magnetic field around a current-bearing solenoid. (Hint: The activity you developed with the paper clips in the follow-up to investigation 6.2 may give you an idea. Do not feel restricted by this hint.) Make sure that your procedure will allow you to answer the follow-up questions.

2. Have your teacher check and approve your procedure. Then carry it out.

3. Test the effect of inserting a core into the solenoid. Try an iron bar, a bar magnet, a copper bar, a piece of wood, a pencil, a straw, and other materials of your choice.

4. Write a report that describes your findings. Include the answers to the follow-up questions.

Follow-up

1. Which core materials, if any, affected the strength of the magnetic field? In what way did they affect it? Did you see an increase or a decrease in strength?

2. If you saw any increase in the magnetic field strength, which core materials gave the greatest increase?

3. Compare your findings with those of other students.

4. How do your findings compare with the prediction you made at the beginning of the investigation?

Investigation	## 6.10 *Methods of Magnetization*

In investigation 6.6, you used a magnetized needle to map lines of force. The needle was already magnetized when you received it. How could it have been magnetized? On the basis of what you have learned about magnets, can you devise and test a method for magnetizing a piece of iron?

Materials

Propose your own materials list when you devise a procedure.

Method

1. Develop your own procedure for magnetizing a piece of iron. Make sure that your procedure will allow you to answer the follow-up questions.

2. Have your teacher check and approve your procedure. Then carry it out.

3. Write a report that describes your findings. Include the answers to the follow-up questions.

Follow-up

1. Suggest at least one other method for magnetizing iron.

2. How did you test the iron to show that you had magnetized it?

3. Once you had magnetized the iron, did your method allow you to make it a stronger magnet? If so, how? How did you try to show that the strength of the magnet was changing?

4. How could you demagnetize the iron once you had magnetized it? Try to demagnetize it. Does your method work?

5. In investigation 6.6, you learned about the internal structure of magnets. Use this knowledge to explain why your methods of magnetization and demagnetization worked. Compare your explanations with those of other students.

6. Compare your methods of magnetization and demagnetization with those of other students. Is it clear which methods are easiest and which are most effective?

| *Investigation* | ## 6.11 The Curie Point |

In investigation 6.10, you explored at least one way of demagnetizing iron. A method you may not have thought of is to heat the iron. As the temperature of a ferromagnetic substance is increased, eventually a temperature is reached at which it ceases to be ferromagnetic. This temperature is called the **Curie point**.

Here are some ferromagnetic elements and their Curie points:

Element	Curie point (°C)
Cobalt	1131
Iron	770
Nickel	358
Gadolinium	16

The following teacher demonstration will allow you to determine the Curie point of an alloy.

An **alloy** is a solid solution, usually containing at least two metals.

Materials

Joe metal (an alloy of 70% Ni, 30% Cu)
small, U-shaped magnet
thermometer
beaker
Bunsen burner
retort stand
3 ring clamps
thread
wire gauze
thermometer clamp
water

Fig. 17

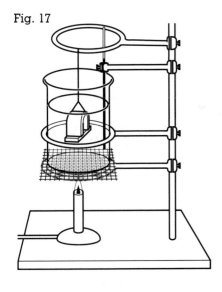

Method

1. Attach a piece of the alloy to the magnet.

2. Suspend the magnet from the ring clamp with thread. Lower the assembly into cold water in the beaker (Figure 17).

3. Gradually heat the water, noting the temperature.

4. Record the temperature at which the alloy drops to the bottom of the beaker. This temperature is the Curie point of the alloy.

Follow-up

1. Can you be sure that the alloy fell to the bottom of the beaker because it lost its ferromagnetic properties? Perhaps the U-shaped magnet lost its magnetic properties. Design an experiment to establish which explanation is correct.

2. From the chart of Curie points, above, would you expect the U-shaped magnet to become demagnetized at the maximum temperature reached by the water in the experiment?

3. Explain, in terms of the internal structure of magnetic materials, why the alloy fell from the magnet at the Curie point.

4. In Chapter 3, you learned about the atomic model of matter. Does your knowledge of the behaviour and structure of magnets and electromagnets support the atomic model? Explain your answer.

Investigation

6.12 Factors Affecting the Magnetic Field of an Electromagnet

In investigations 6.7 and 6.9, you began to examine electromagnets. You have already learned that changing the material in the core of a solenoid can change the strength of the magnetic field. But what other factors affect the field strength of an electromagnet? Think of the solenoid you built earlier, and suggest variables that might affect its field strength. Note and discuss your ideas.

Materials

Propose your own materials list when you devise a procedure.

Method

1. Develop your own procedure for testing your hypotheses. Remember the importance of a controlled experiment. In other words, if you want examine several variables, change them only one at a time. Make sure that your procedure will allow you to answer the follow-up questions.

2. Have your teacher check and approve your procedure. Then carry it out.

3. Write a report that describes your findings. Include the answers to the follow-up questions.

Follow-up

1. What factors affect the field strength of the solenoid?

2. Compare your findings with those of other students.

3. When you complete your investigation, discuss with your teacher the various factors that affect the field strength. Did your procedure reveal them all? If not, how could you have modified your procedure to investigate the variables you overlooked?

| Investigation | # 6.13 A Quantitative Look at Field Strength |

In the previous investigation, you discovered the factors that affect the strength of an electromagnet. However, the results of the investigation were qualitative. You can now use the following procedure to determine the quantitative dependence of the strength of an electromagnet on two factors that control it.

Materials

copper wire
large, thick nail
power supply
connecting wires
paper clips

Method

1. Wrap copper wire around a core (the nail) and tape it in place to create a solenoid.

2. Connect the ends of the wire to a power supply. Pass 1.0 A of current through the solenoid.

3. Test the strength of the solenoid by counting the number of paper clips it can pick up at one time.

4. Repeat step 3, but increase the current to 2.0 A, then 3.0 A, 4.0 A, and finally 5.0 A.

5. Adjust the solenoid so that there are 10 loops of wire around the nail. (Spread the loops along the entire nail, rather than bunching them together.)

6. Again, connect the ends of the wire to the power supply. Adjust the current to 1.0 A and count the number of paper clips the solenoid can lift.

7. Repeat step 6, but with 15, 20, 25, and finally 30 loops of wire around the nail. Make sure that the length of the solenoid stays the same in each case by spreading the loops along the length of the nail. Keep the current at 1.0 A.

A is the symbol for the SI unit of electric current, the ampere.

Follow-up

1. Plot graphs to show the relationship of the strength of an electromagnet (as shown by the number of paper clips it lifts) and:
   (a) the current through the electromagnet
   (b) the number of loops of wire

2. What kind of graph do you obtain in each case in question 1?

3. What kind of relationship does each graph suggest?

4. Formulate a mathematical relationship between the strength, B, of an electromagnet and both the number of loops, n, and the current, I.

5. Compare your formula from question 4 with those of other students.

6. The SI unit of the field strength is the tesla, symbol T. Find out how the tesla is defined and discuss the definition.

7. A solenoid can pick up six paper clips. If the current through the solenoid is halved and the number of loops is tripled, how many paper clips do you expect the solenoid to lift?

8. Write a report on the investigation and include your answers to the follow-up questions.

6.14 Magnetism and Consumer Goods

You have now learned a good deal about the principles of magnetism and electromagnetism. Investigation 6.2 included some information about the applications of magnets. It is important to realize the extent of these applications, their impact on society, and how they relate to scientific principles. The following questions will encourage you to research and think about these aspects of magnetism and electromagnetism.

Questions

1. What consumer goods are made through the application of magnetic and electromagnetic principles?

2. What is the role of magnetism or electromagnetism in each of the consumer goods you listed?

ELECTROMAGNETISM AND HUMAN HEALTH

Some scientists have expressed concerns that electromagnetic fields produced by the flow of electric current through power lines and household appliances might be dangerous. Could electric blankets, for example, be harmful to people?

Studies carried out in the United States in the past two decades have suggested a link between exposure to electromagnetic fields and high rates of brain cancer and leukemia in children. However, the news about electromagnetic fields may not be all bad. Klaus-Peter Ossenkopp of the University of Western Ontario has found a beneficial effect on rats with epilepsy.

3. What is the impact of the application of magnetic and electromagnetic principles in the manufacture of consumer goods? Think about:
(a) the economic impact   (b) the environmental impact

4. Research is going on into the possible effects of magnetism and electromagnetism on human health. Find out and discuss the latest results of this research.

Fig. 18 *Electromagnets are used to separate metals in scrap yards.*

6.15 Magnetism—Past, Present, and Future

In section 6.14, you considered some modern-day applications of magnetism and electromagnetism. It is also interesting to consider the past and the future.

Write a short paper on the discovery and applications of the principles of magnetism and electromagnetism. In discussing this topic, you should:

1. Identify the scientists who developed these principles.

2. Describe the effects of magnetism and electromagnetism on human activities at different times in history.

3. Identify areas of advanced research in magnetism and electromagnetism.

4. Develop ideas on some possible future applications of magnetic and electromagnetic principles.

To check whether the fears about electromagnetic fields are justified, Ontario Hydro, Hydro-Québec, and Health and Welfare Canada began a three-year study in 1990. They will attempt to determine the effects of electromagnetic fields on the development of brain cancer and leukemia in laboratory animals.

Fig. 19 *Professor Klaus-Peter Ossenkopp.*

P O I N T S · T O · R E C A L L

- Certain materials, such as lodestone, exert magnetic forces. We call these materials magnets.
- A magnet has two poles, north and south.
- A magnetic field is the region around a magnet where magnetic forces of attraction and repulsion are felt.
- The line along which a magnet moves a free north pole is called a line of force.
- The Earth has a magnetic field similar to that of a bar magnet.
- The Earth's magnetic poles do not coincide with its geographic poles.
- Substances that contain certain elements, such as iron, cobalt, and nickel, are attracted by magnets and can be magnetized. They are known as ferromagnetic substances.
- There are various ways of magnetizing and demagnetizing ferromagnetic substances.
- A material's ability to stay magnetized is called remanence or retentivity.
- Magnetism may be explained by means of the theory of domains.
- Whenever an electric current moves through a conductor, a magnetic field is created.
- The minimum temperature at which a ferromagnetic substance is no longer attracted by a magnet is called the Curie point.
- The magnetic field strength of a solenoid depends on several factors.
- Magnetism and electromagnetism have many important applications.

R E V I E W · Q U E S T I O N S

1. Name:
   (a) three ferromagnetic substances
   (b) three non-magnetic substances
2. What name is given to the region around a magnet where magnetic effects are felt?
3. Why do surveyors in most parts of the world make corrections to their compass readings?
4. State the laws of magnetic forces.
5. Explain what is meant by:
   (a) domain   (b) remanence
6. Use the domain theory to explain the difference in magnetization of "soft" iron and "hard" steel. Which would you choose as a compass needle and why?
7. Explain in terms of domains how the following affect the strength of a magnet:
   (a) dropping the magnet
   (b) strongly heating the magnet
8. What is the Curie point?
9. A paper clip is magnetized and is then straightened out. How many poles does it have before and after being straightened?
10. State three characteristics of magnetic lines of force.
11. What factors affect the strength of an electromagnet?
12. If the N-pole of a magnet attracts the S-pole of another magnet, is the Earth's magnetic north pole really a north pole? Explain your reasoning.
13. If a steel ship is built with its bow pointing toward the Earth's magnetic north pole, the hammering during construction may result in the ship being magnetized. On the basis of your answer to question 12, decide which end of the ship will have north polarity. Explain your reasoning.
14. To make a watch that is protected against the effects of magnets, should you make its casing from magnetic, ferromagnetic, or non-magnetic material? Explain.

15. What is the difference between a temporary and a permanent magnet?
16. Three iron bars look identical, but two are magnets and one is not. Explain how you would distinguish the bars without using any additional materials.
17. Explain how you would determine which end of a magnet is the N-pole.
18. Suppose that you dropped a cassette tape on the floor five or six times over a short period. Would the quality of the recording stay the same? Explain.
19. An electromagnet is found to be too weak for the purpose intended. How can you increase its strength?
20. Consider a solenoid having 10 loops of wire carrying 2.0 A, and a similar solenoid having 20 loops of wire carrying 4.0 A.
    (a) Which solenoid exerts the greater magnetic force?
    (b) By what factor do the forces differ?
21. An electromagnet is made of 500 loops of wire. An electric current of 3.0 A creates a certain lifting force. What current is required to provide twice as great a force if the number of loops is doubled?
22. Figure 20 shows an electromagnet. Which pole is north and which is south?

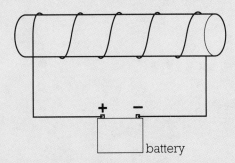

Fig. 20

7

CURRENT ELECTRICITY AND ELECTRIC CIRCUITS

High-voltage electricity is hazardous. Always handle electrical equipment with caution.

CONTENTS

7.1 Electricity at Rest and in Motion

Ebonite is a hard, black material made by heating rubber with sulfur.

Fig. 1 *Lightning is a discharge of static electricity.*

The Greek philosopher Thales knew that amber, when rubbed, attracted very light objects. No doubt, you have rubbed a plastic comb or pen and used it to pick up small pieces of paper. Many other substances, including ebonite and glass, behave in this way when rubbed. As you learned in investigation 3.4 and section 3.5, the rubbing process results in a buildup of static electricity.

You meet many examples of static electricity every day. A nylon garment may crackle when you take it off. The crackles are caused by tiny electric sparks. You may hear similar noises when you separate clothes that have been in the dryer. But sparks from static electricity are not always small. A much bigger example is a bolt of lightning (Figure 1).

A spark shows that a charge is jumping from one place to another. This rapid transfer of charge, which removes the static electricity from an object, is called a **discharge**. But the discharge of static electricity is not the only way that electric charges can get from one place to another. In this chapter, we will investigate electricity that travels through wires in electric circuits. This kind of electricity is called **current electricity**.

167

| *Investigation* | ## 7.2 *Conductors and Insulators* |

Some substances readily allow electricity to pass through them, while others do not. Substances that carry electricity well are called **conductors**. Substances that do not readily allow electricity to flow are known as **insulators**. How well a material conducts is one of its characteristic properties.

There is no sharp line that divides conductors from insulators. Most insulators conduct a little, and even good conductors vary greatly in their ability to conduct. So, how easy is it to classify materials on the basis of how well they conduct?

To answer this question, sort the following materials into two or three groups, on the basis of how well they conduct electricity. You will need the symbols you used in an earlier course to represent parts of an electric circuit. Recall that in Figure 2, below, is an example of a "circuit diagram." If you are unsure about how to interpret the symbols in Figure 2, then refer to Figure 6 in investigation 7.4.

Materials

glass rod
copper metal
wooden splint
piece of plastic or rubber
aluminum metal
graphite
piece of ceramic (such as pottery)
power source
flashlight bulb
connecting wires

Method

1. Construct the circuit as shown in Figure 2.

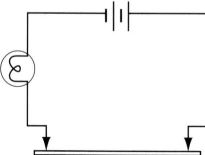

Fig. 2

2. Test each substance. Make up your own system for describing the brightness of the light in each case.

3. Classify the materials according to their ability to conduct electricity.

Follow-up

1. Was your way the only possible way of:
   (a) describing brightness?
   (b) classifying materials on the basis of how well they conduct?

   Discuss this question with your group and consult your teacher.

2. If you were asked to design an electrical plug, what properties should the prongs and casing have?

3. Most electrical tools have handles covered with heavy rubber. Explain why.

4. Why are wooden poles used to carry electrical power lines in your neighbourhood?

5. You cannot seem to charge a copper rod by rubbing it, no matter how hard you try. However, if you fit the rod with a plastic handle, you can build up a charge. Explain why.

6. Think about the various components in the electric circuit you set up. Can you classify them on the basis of their ability to conduct? (For example, think about the electrical wires. What are they made of and why? What material makes up the covering on the wires, and why is this material chosen?)

7. What are some of the commonly used conductors and insulators? (For example, what materials are used in household wiring, what materials are used as insulators in power stations, and so on?) Check reference sources and draw up a list. Discuss your findings.

| Investigation | 7.3 *Conductance of an Object* |

The ability of an object to conduct electricity is known as its **conductance**. Our lives would be very different if we did not have objects that conduct well (Figure 3). But what are the factors that affect the conductance of an object? Decide what you think these factors might be and discuss your ideas.

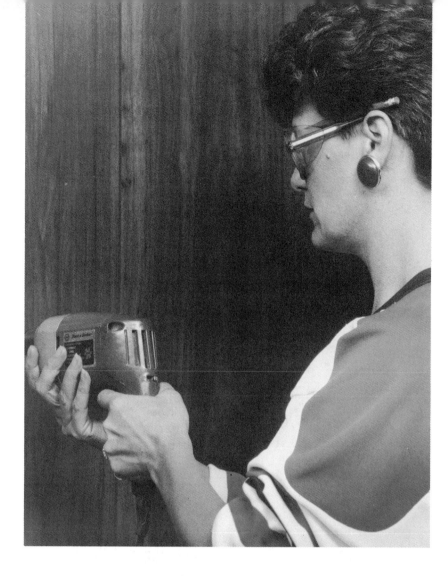

Fig. 3 *Electrical applications depend on wires that conduct.*

Do not carry out any procedure of your own until your teacher has checked it for safety.

Materials

Propose your own materials list when you devise a procedure.

Method

1. Design a simple circuit diagram to help you identify the variables that determine conductance.

2. Outline a procedure to test your hypotheses. (If you think that more than one variable affects the measured quantity, remember to test the effect of one variable at a time. Keep the other variables constant.) Make sure that your procedure will allow you to answer the follow-up questions.

3. Consult your teacher to have your procedure approved or modified.

4. Perform the experiment.

5. Submit your results for class discussion.

6. Write a report using class data.

Follow-up

1. When you compared wires with the same dimensions, but made of different materials, were their conductances the same or different?

2. Does increasing the length of a wire have any effect on conductance? If so, does conductance go up or down as the length increases?

3. Does the cross-sectional area of a wire have any effect on conductance. If so, does conductance go up or down as the cross-sectional area increases?

Follow safety rules while heating or cooling objects.

4. In investigation 6.11, you learned that temperature affects the alignment of magnetic domains. Do you think that temperature might also affect conductance? Explain why you think temperature will or will not affect conductance.

5. Test your hypothesis about the effect of temperature. Design an investigation, have it approved by your teacher, and carry it out. What is the effect of temperature, if any?

6. Did you test other variables for their effect on conductance? If so, what were your findings?

7. From your answers to questions 2, 3, 5, and 6, list the factors that affect the conductance of an object.

8. Do the generalizations you reached in question 7 apply to all materials? Research this point in reference books.

9. Why are electrical transmission cables thick?

10. If you were wiring a house, would you try to keep the connecting wires as short as possible, or would length not matter? Explain your answer.

11. Do outdoor electrical cables conduct better in winter or in summer, or does the season make no difference? (Assume that the dimensions of the cables stay roughly constant.)

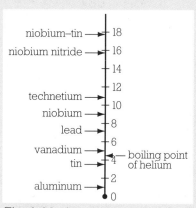

ELECTRICAL TIDBITS

Electrical resistance is the opposition to the flow of electricity. So, the lower the resistance of an object, the higher its conductance.

niobium–tin → 18
niobium nitride → 16
— 14
— 12
technetium → 10
niobium →
lead → 8
— 6
vanadium →
tin → 4 — boiling point of helium
— 2
aluminum →
0

Fig. 4 *Maximum temperatures (in kelvin) at which these materials are superconductors.*

SUPERCONDUCTIVITY

At very low temperatures, materials behave very differently than we are used to. For example, it has been found that very low temperatures can have a dramatic effect on how well substances conduct.

In 1911, H. Kamerlingh Onnes, at the University of Leiden in the Netherlands, cooled mercury to a lower temperature than anyone had before. At the boiling point of liquid helium (4.2 K or −269°C), the frozen mercury became an unbelievably good conductor. Physicists describe frozen mercury at these temperatures as having zero "resistance" to the passage of electricity.

Another experiment that Onnes carried out involved cooling a coil of lead in liquid helium. When he passed an electric current through the coil, it produced a magnetic field, as you would expect. But, after he turned the current off, the magnetic field was still there. In other words, the current was still flowing through the coil without the help of the battery!

Such amazing conductors are called **superconductors**. A number of metals and their compounds, as well as some metal alloys, behave as superconductors when they are cold enough. Some examples are shown in Figure 4.

The world's most powerful magnets are made of coils of superconducting wires. An enormous current flows in the coils. Some of these magnets produce fields that are 200 000 times greater than the field of the Earth at its magnetic poles. Such powerful magnets have made possible "magnetic imaging," used in medical diagnosis.

| *Investigation* | ## 7.4 *Current Intensity in a Circuit Element* |

We will now begin to examine current electricity in greater detail. An important skill in designing electric circuits is the ability to represent them in a shorthand way. Figure 2 in investigation 7.2 is an example of a shorthand diagram, known as a circuit diagram. We use a set of conventional symbols in circuit diagrams, which you learned in a previous course. Refer to Figure 6 for a quick review.

There are many other possible uses of powerful magnets. One is the magnetic levitation (maglev) train. There is already a small maglev train in Japan. It floats along above its pathway, supported by magnetic repulsion. As there is no friction with the tracks, it is much faster than other trains. It cruises at 640 km/h!

No heat is produced when a current flows through a superconductor, so, in the future, large currents may be sent long distances without loss of electrical energy in the form of heat. Also, superconductors may one day be used to store large quantitities of electrical energy. Electricity could be pumped into a circular superconductor and be left to go round and round until the energy is needed.

As research continues, materials are being developed that become superconductors at higher and higher temperatures. So the days of widespread applications seem not too far off. Of course, materials that superconduct at room temperature or above would be ideal. They would not have to be cooled. Perhaps some day we will have such superconductors.

Before we leave the subject of superconductors, you should realize that they are not just "normal" conductors that conduct better at lower temperatures. Superconductors are very different from the conductors you are familiar with. In fact, the best of our conventional conductors, such as silver and copper, do not become superconductors at low temperatures. On the other hand, some recently discovered superconductors are ceramic materials. These act as insulators under normal conditions.

Fig. 5 *An experimental maglev train in Japan.*

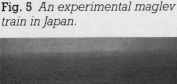

cell

battery

resistor

lamp

ammeter

voltmeter

Fig. 6 *Some conventional symbols used in circuit diagrams.*

galvanometer

switch

173

Fig. 7 *A flowmeter and an ammeter are connected in similar ways.*

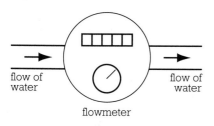

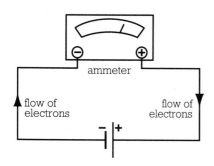

An important feature of current electricity is the **electric current**, sometimes called the "current intensity" or just the "current." It is usually represented by the letter *I*. The SI unit of electric current is the ampere, symbol A.

But what exactly is an electric current and how do we measure it? We need an important piece of equipment called an ammeter. This device measures the flow of current through a conductor, somewhat like a flowmeter measures the flow of water through a pipe. In both cases, the meter is inserted into the "circuit" in such a way that all the water or electric current flows through it (Figure 7).

Both the flowmeter and the ammeter measure the flow in one direction. So it is important that they be connected the right way around. In the case of the ammeter, the negative terminal, marked (−), is connected to the negative terminal, (−), of the power supply. The positive terminals, (+), of the ammeter and power supply must also be connected. Your teacher will give you more details of how to use the ammeter without damaging it.

Materials

power supply
switch
flashlight bulb
other circuit elements (resistors)
connecting wires
ammeter

Method

1. Draw a circuit diagram to represent a circuit that connects the power supply, switch, and light bulb.

2. Assemble the circuit you drew in question 1 and close the switch.

3. Now that the current is flowing, think about what an electric current is. Discuss your concept of an electric current with your group.

4. Connect the ammeter into your circuit in such a way that the entire circuit is just one loop. Physicists describe this kind of connection as a **series connection**. Another way of describing the assembly is to say that everything in the circuit is connected "in series." Record the reading on the ammeter.

5. Draw the circuit diagram for the circuit with the ammeter in place. Use the conventional symbol for an ammeter (see Figure 6). On the circuit diagram, indicate the measured current

intensity. (Say, for example, your measured value is 0.22 A. A common way to show this on the diagram is to write "I = 0.22 A" parallel to the symbol for a connecting wire.)

.ecord the correct number of .gures, such that the final one s estimated. Add the unit.

6. Replace the light bulb with another circuit element. Again, record the reading on the ammeter, draw a circuit diagram, and mark on it the current intensity.

7. Repeat the process for the other circuit elements supplied.

Follow-up

1. What role does the power supply play in a circuit?

2. What condition(s) must be met for a current to flow in a circuit?

3. Suppose someone claimed that "the current intensity in a circuit should only depend on the power supply you use." Discuss this statement in the light of your results.

4. Suppose that you were able to look inside a connecting wire in a circuit to watch the particles. What would you see happening when the current flows? If the current intensity increased, what change would you see?

ELECTRICAL SAFETY

Too great a current intensity may make conductors so hot that they burn off their insulation and start fires. For this reason, each circuit in a building has a fuse or a circuit breaker. It acts as an emergency switch. It opens the circuit automatically if the current intensity is too great.

For a fuse or circuit breaker to work properly, it must match the circuit it is part of. For example, a 15-A household circuit should have a 15-A fuse or circuit breaker. Some people take great risks by putting, say, a 30-A fuse in a 15-A circuit. Then, if the current intensity goes over 15 A, the fuse may not "blow," and the fire risk increases.

As an independent reading exercise, find out how fuses and circuit breakers work.

Fig. 8 *Every home has a fuse box or circuit breaker panel.*

| # 7.5 Potential Difference in a Circuit

Water flows through a pipe if there is a pressure created by different water levels at the start and end of the pipe, or if a pump pushes the water along. Might the flow of electrons through a circuit resemble water flow in any way(s)? What does make electricity flow through a circuit? An important idea in this context is **potential difference**, usually represented by the letter V.

You may often hear potential difference referred to as "voltage."

You will recall the term "potential difference" from a previous course. But how is potential difference measured? The instrument we use is called a voltmeter. In this investigation, you will learn how to use it. You need to know that the SI unit of potential difference is the volt, symbol V.

V for the variable (potential difference) is in italics. V for the unit (volt) is not in italics.

Another important point is that a voltmeter is not connected in series in a circuit. Instead it is connected **in parallel**. This means that the voltmeter is on its own loop. The ends of this loop are connected to two points in the circuit (Figure 9). The voltmeter measures the potential difference across the part of the circuit between these two connection points.

You can think of a voltmeter as a device that measures potential difference, somewhat like a pressure gauge measures water pressure in a pipe. Both measuring instruments are connected in parallel (Figure 10).

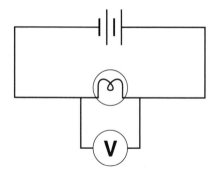

Fig. 9 *The voltmeter is connected in parallel with the lamp.*

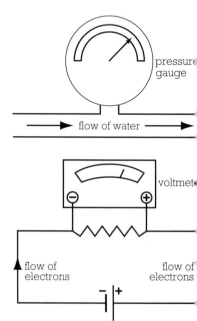

Fig. 10 *A pressure gauge and a voltmeter are connected in similar ways.*

In each case, the meter must allow very little current (flowing water or flowing electricity) to pass through it. Otherwise, it will give a false reading by disturbing the flow in the rest of the circuit too much. In electrical terminology, a voltmeter must have a high resistance, so that most of the current flows through the rest of the circuit and not through the meter. This is very different from the requirements of an ammeter. It measures the current intensity, so the current must go through it. Therefore, the ammeter is connected in series, not in parallel. Because the current must flow through an ammeter, its resistance must be very low. Otherwise, it will interfere with the flow too much and give a false reading of the current intensity.

The direction of flow through the meter is important. For the voltmeter, make sure that the negative terminal, ($-$), is connected to the negative terminal, ($-$), of the battery. The two positive terminals, ($+$), must also be connected. Your teacher will give you more details of how to use the voltmeter without damaging it.

Materials

power supply
switch
flashlight bulbs of different voltages
connecting wires
voltmeter

Method

1. Draw a circuit diagram to represent a circuit that connects the power supply, the switch, and a light bulb.

2. Assemble the circuit you drew in question 1.

3. Connect the voltmeter to your circuit, such that it records the potential difference across the lightbulb. Record the reading on the voltmeter.

4. Draw the circuit diagram for the circuit with the voltmeter in place. Use the conventional symbol for a voltmeter (Figure 6). On the circuit diagram, indicate the measured potential difference. (Say, for example, your measured value is 0.90 V. A common way to show this on the diagram is to write "$V = 0.90$ V" next to the symbol for the voltmeter.)

5. Replace the light bulb with a different one. Again, record the reading on the voltmeter, draw a circuit diagram, and mark on it the potential difference across the bulb.

6. Repeat the process for the other bulbs supplied.

Follow-up

1. What creates the potential difference across each circuit element

2. Suppose that the potential difference of a battery is 9.0 V. Does this value indicate the size of a force, a quantity of energy, or both?

3. Having connected a voltmeter to a circuit, you cannot get a reading. Upon closer inspection, you notice that the needle is to the left of the zero mark on the scale.
   (a) What is wrong?
   (b) How can you correct the problem?

4. What would happen if you connected a voltmeter with a maximum reading of 100 V to each of the following?
   (a) a 220-V line
   (b) a 50-V line

5. Voltmeters and ammeters look much the same. What would happen if you made a mistake and connected a voltmeter instead of an ammeter in series with a lamp?

Investigation	*7.6 Electrical Conductance*

In investigation 7.3, you examined the factors that affect the electrical conductance of an object. Unless you decided to examine the effect of changing the power supply, you have yet to see how a change in the potential difference might affect the current intensity in a circuit element. What do you think the effect, if any, of changing the potential difference will be? Whether you think that changing the potential difference will affect the current intensity or not, explain your reasoning. Discuss your prediction and explanation with your group.

Materials

Propose your own materials list when you devise a procedure.

Method

1. Develop a procedure for measuring the current intensity in a flashlight bulb when the potential difference varies. Think about how to vary the potential difference. What kind of source do you need?

2. In your design for the circuit, apply your earlier knowledge of how to connect an ammeter and voltmeter.

3. Have your teacher check your circuit before you begin to take measurements.

4. Conduct the experiment and organize your data into a chart to find relationships between the measurements.

To graph *I* against *V*, put *I* on the vertical axis and *V* on the horizontal axis.

5. Plot a graph of current intensity, *I*, against potential difference, *V*. Determine the slope of the line you obtain and write this value on the line. (If you are unsure about how to determine the slope, refer to Appendix 2.)

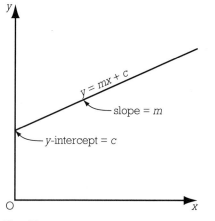

Fig. 11

Follow-up

1. Calculate the ratio $\dfrac{\text{current intensity}}{\text{potential difference}}$ for every pair of measurements you recorded. The value you obtain is the proportionality constant that relates current intensity to potential difference. Compare this proportionality constant with the slope of the line.

2. What is the mathematical relationship of the current intensity, the slope, and the potential difference? (Recall that the equation of a line is of the form $y = mx + c$, where y is the variable on the vertical axis of the graph, x is the variable on the horizontal axis, and m is the slope of the line. The constant term, c, is zero for any line that passes through the origin of the axes. If you need more help in relating the graph to the equation of a line, refer to Appendix 2.)

3. The proportionality constant and the slope of the line both equal the conductance of the circuit element. In other words, conductance, G, is the ratio of current intensity, I, to potential difference, V. Thus, we can write the equation $G = \dfrac{I}{V}$. On the basis of this definition, what units do you think conductance is expressed in? (The unit you predict is not the standard one. The standard unit of conductance is the siemens, symbol S.)

4. The relationship between the current intensity and potential difference for a circuit element was first discovered by a famous German mathematician and physicist, Georg Simon Ohm. Look up and state the law that he developed.

Fig. 12 *Georg Simon Ohm.*

179

ELECTRICAL
TIDBITS

ELECTRIC SHOCK

Electricity is so common that sometimes we become careless with it. Such carelessness may result in shocks or, in extreme cases, in fatal electrocution.

In investigation 7.5, you learned that the current intensity in a conductor depends on both its conductance and the applied potential difference. The conductance of an object can greatly increase when the object is directly connected to the Earth, in other words, when the object is **grounded**. A small voltage may then be sufficient to send a large current through the object. This explains why touching an electrical device, such as a lamp or radio, can electrocute people in their bathtubs.

© 1991 Tribune Media Services, Inc.

Why can birds safely perch on a high-voltage line, while you get electrocuted if you touch the same wire? The bird has both feet on the line, so there is practically no difference in electric potential between the bird and the line. Hence, the current passing through the bird is negligible. However, if you were to touch the line (either directly or with a conductor, such as a metal ladder) while your feet were on the ground, your conductance would increase. The large potential difference between the line and the ground would send a very large current through you.

If an electric current through the human body exceeds 0.01 A, people lose voluntary muscle control. So, if they are holding a live wire, they cannot let go of it. They may have violent muscular contractions and cramps in their arms. A current above 0.1 A stops the heart from beating normally.

| *Investigation* | *7.7 Conductance and Resistance* |

You have learned some very general definitions of conductance and resistance. Conductance is the ability of an object to carry an electric current, whereas resistance is the opposition an object shows to the passage of electricity. These ideas sound like opposites, but what exactly is the relationship between them?

Before you begin numerical work, think about conduction of electricity at the atomic level inside a conductor. Why do you think that conductors have electrical resistance? Discuss your ideas with your group and with your teacher.

Materials

Method

1. Consult the current intensity and potential difference data you recorded in the previous investigation.

2. Again, plot a graph, but reverse the axes (i.e., plot V against I). Determine the slope of the line and record its value on the line. This slope gives the resistance, R, of the circuit element.

Follow-up

1. Recall that the slope you determined in the previous investigation was the conductance of the circuit element. What is the mathematical relationship between resistance, R, and conductance, G?

2. By considering a ratio or the equation of the line you plotted, derive a mathematical formula that relates resistance, R, to potential difference, V, and current intensity, I.

3. From your graph, predict the units of electrical resistance. (Your prediction will not be the standard unit, which is the ohm, symbol Ω.)

Pronounce "ohm" to rhyme with "home."

4. In investigation 7.3, you discovered the factors that affect the conductance of an object. Now that you know the relationship between conductance and resistance, decide how each of those factors affects the resistance of an object. Explain your reasoning.

7.8 Numerical Problems Involving R, V, and I

Now that you know the relationship between resistance, potential difference, and current intensity, you are in a position to complete some numerical problems that include these variables.

Example 1

An electric lamp uses a current of 1.0 A when the potential difference across it is 110 V. What is the resistance of the lamp?

Solution

$$R = \frac{V}{I}$$

$$= \frac{110 \text{ V}}{1.0 \text{ A}}$$

$$= 1.1 \times 10^2 \ \Omega$$

Example 2

What is the current intensity in a 5.5-Ω heater when it is connected to a 110-V household circuit?

Solution

$$R = \frac{V}{I}$$

$$\text{So} \quad I = \frac{V}{R}$$

$$= \frac{110 \text{ V}}{5.5 \ \Omega}$$

$$= 2.0 \times 10^1 \text{ A}$$

Example 3

What potential difference is needed to push 5.0 A of current through a 4.0-Ω lamp?

Solution

$$R = \frac{V}{I}$$

So $V = IR$

$$= 5.0 \text{ A} \times 4.0 \text{ } \Omega$$

$$= 2.0 \times 10^1 \text{ V}$$

Note that, whenever you do a calculation, it is a good idea to check that your answer is reasonable. For instance, in Example 2, above, the calculation involves the division of a number close to 100 by a number close to 5, so an answer of 20 is reasonable. Without some kind of mental check, it can be dangerous to believe the answer you see on a calculator screen. Anyone can input a wrong number or press a wrong operator button.

Check also that the units of your answers make sense. If you are using standard SI units in calculations that involve the equation $V = IR$, then V is in volts, I in amperes, and R in ohms.

Questions

1. Calculate the current through a 24-Ω resistor connected to a 12-V battery.

2. A current of 6.0 A flows through an electric iron connected to a 120-V source. What is the resistance of the electric iron?

3. A toaster intended for use on a 110-V circuit has a resistance of 27.5 Ω. What is the current intensity through the toaster?

| *Investigation* | *7.9 Codes for Resistors* |

Anything with electrical resistance is called a **resistor**. However, this term is often used in a narrower sense to describe a certain kind of device. Such a device is usually made of carbon or high-resistance wire and may have a variable or fixed resistance. Some types of resistors have a ceramic coating with a colour code to indicate the resistance.

You now know the relationship between resistance, current intensity, and potential difference. One way to vary the current in a circuit is to vary its resistance. So, resistors help us to design electrical circuits for particular purposes.

Fig. 13 *Resistor colour codes.*

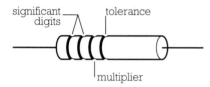

You will now examine some colour-coded resistors to determine their resistances. You first need to know what the four coloured bands signify (Figure 13).

Resistor Colour Codes			
Colour	**Digit**	**Multiplier**	**Tolerance**
black	0	10^0 or 1	
brown	1	10^1	
red	2	10^2	
orange	3	10^3	
yellow	4	10^4	
green	5	10^5	
blue	6	10^6	
violet	7	10^7	
gray	8	10^8	
white	9	10^9	
gold		10^{-1}	$\pm 5\%$
silver		10^{-2}	$\pm 10\%$
no colour			$\pm 20\%$

The reference chart will allow you to interpret the coloured bands. For example, suppose that the bands in order are red, brown, orange, and gold, beginning with bands that represent the significant digits. The red and brown bands indicate the number 21. The orange band gives a multiplier of 10^3. Therefore, the resistance is $21 \times 10^3 \ \Omega$, or $21\ 000 \ \Omega$. The gold band, which shows the "tolerance," indicates that the actual resistance should be within 5% of the calculated value.

Materials

assortment of colour-coded resistors

Method

1. Obtain some colour-coded resistors from your teacher.

2. Consult the reference chart and determine the resistance of each resistor.

Follow-up

1. Suppose that a simple circuit contains just a power supply, a resistor, and two connecting wires. If you replace the resistor with one of lower resistance, and you keep the potential difference constant, what happens to the current intensity?

2. The coloured bands on a resistor are, in order (beginning with the significant digit bands), yellow, black, green, and silver. Determine the resistance and tolerance of this resistor.

3. If, to a tolerance of $\pm 5\%$, the resistance of a resistor is 76 000 000 Ω, what bands should appear on the resistor?

| Investigation | # 7.10 Equivalent Resistance in Series and Parallel Circuits |

You have examined the relationship between the current intensity, resistance, and potential difference for a single resistor, such as a light bulb, in a circuit. However, many circuits are more complicated than this. What if a circuit has two or more resistors? Does it make any difference to the variables if the resistors are connected in series or in parallel? For example, suppose you had two identical light bulbs connected to a battery. Does the overall resistance of the circuit depend on whether the bulbs are in series or in parallel? Suggest what you think the answer might be and discuss it with your group. The following activity will allow you to test your ideas.

Before you begin, you should learn the meaning of an important term, **equivalent resistance**. It is defined as the resistance of a single resistor that could replace two or more series or parallel resistors without changing the total current in the circuit. This is a fairly difficult definition. An example will help clarify it. Suppose you have a circuit with two resistors of resistance x Ω and y Ω connected in series. You find that the current in the circuit stays the same if you replace the two resistors with a single resistor of resistance z Ω (Figure 14). Then you can say that z Ω is the equivalent resistance of the two resistors you replaced. You can repeat the procedure, beginning with the x Ω and y Ω resistors in parallel, to find out whether their equivalent resistance is different when they are connected differently.

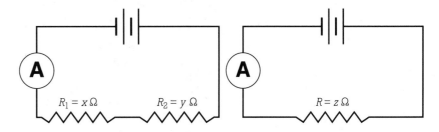

Fig. 14 *If the current stays the same, R is the equivalent resistance of resistors R_1 and R_2.*

$R_1 = x\,\Omega$ $R_2 = y\,\Omega$ $R = z\,\Omega$

185

Materials

Develop your own materials list as you devise a procedure.

Method

I. *Series circuit*

1. Develop a procedure to determine the equivalent resistance of a circuit that contains at least two resistors in series. Use your previous knowledge of the ammeter and voltmeter, and remember that resistance is the slope of the line obtained from a graph of potential difference against current intensity.

2. Draw a circuit diagram and have it checked and approved by your teacher.

3. Carry out the experiment. Tabulate and plot the data to obtain the equivalent resistance of the circuit.

4. Check the value you obtained in step 3 by measuring the equivalent resistance with an ohmmeter. (Your teacher will show you how to use it.)

5. Suggest possible explanations for any discrepancies between the values obtained in steps 3 and 4.

II. *Parallel Circuit*

1. Repeat steps 1 to 5 for a circuit that contains at least two resistors in parallel.

Follow-up

1. Suggest a mathematical relationship between the equivalent resistance of the circuit and the resistances of individual resistors for:
   (a) a series circuit   (b) a parallel circuit

   (The relationship is quite difficult to find in (b). You may need help from your group and from your teacher.)

2. Compare your suggestions from question 1 with those of other students.

3. Check your relationships by using them to calculate the equivalent resistance of each circuit from the resistances of the individual resistors.

4. Compare your calculated equivalent resistances to the values you obtained in steps 3 and 4 of the procedure for each circuit. Suggest reasons for any differences.

5. Use your mathematical relationships to calculate the equivalent resistance of each of the following circuits:

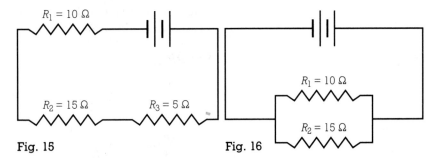

Fig. 15   Fig. 16

6. Find out and list some examples of series and parallel circuits that affect your life in some way.

7. Write a report that includes your procedure, data, graphs, and responses to the follow-up section.

7.11 Numerical Problems on Equivalent Resistance

You can now use the relationships derived from the previous investigation in numerical problems.

Example 1

What is the equivalent resistance of a light bulb of 4.0 Ω, a toy motor of 3.0 Ω, and an electric buzzer of 12.0 Ω, if you connect them in:
(a) series?
(b) parallel?

Solution

(a) Let the individual resistances be R_1, R_2, and R_3.

The equivalent resistance, $R = R_1 + R_2 + R_3$
$$= 4.0\ \Omega + 3.0\ \Omega + 12.0\ \Omega$$
$$= 19.0\ \Omega$$

(b) $\dfrac{1}{R} = \dfrac{1}{R_1} + \dfrac{1}{R_2} + \dfrac{1}{R_3}$

$\dfrac{1}{R} = \dfrac{1}{4.0} + \dfrac{1}{3.0} + \dfrac{1}{12.0}$

$R = 1.5\ \Omega$

187

Example 2

A TV and a chandelier are connected in parallel. If their equivalent resistance is 4 Ω and the resistance of the TV is 12 Ω, what is the resistance of the chandelier?

Solution

$$\frac{1}{R} = \frac{1}{R_1} + \frac{1}{R_2}$$

$$\frac{1}{4} = \frac{1}{12} + \frac{1}{R_2}$$

$$\frac{1}{4} - \frac{1}{12} = \frac{1}{R_2}$$

$$R_2 = 6 \ \Omega$$

The resistance of the chandelier is 6 Ω.

Questions

1. Your stereo system may have a speaker output circuit designed for 6-Ω speakers. What is the equivalent resistance of your speakers if you connect three of them to your stereo:
   (a) in series?
   (b) in parallel?

2. What is the equivalent resistance of five 4-Ω light bulbs connected:
   (a) in series?
   (b) in parallel?

3. Three resistors connected in parallel have an equivalent resistance of 2 Ω. Two of the resistors have resistances of 5 Ω and 4 Ω, respectively. What is the resistance of the third resistor?

4. You decide to replace a defective 2-Ω resistor in your TV set. You do not have a 2-Ω resistor to replace it with, but you do have a supply of 4-Ω and 8-Ω resistors. Determine how you could connect any number of them to obtain an equivalent resistance of 2 Ω.

5. If you have two resistors you wish to connect in such a way that their equivalent resistance is as low as possible, should you connect them in series or in parallel?

6. Because the equivalent resistance of resistors in series is obtained by addition, it must be true that the equivalent resistance is greater than any individual resistance in the circuit. Now consider the statement: "The equivalent resistance of at least two resistors in parallel is always lower than the resistance of any of the individual resistors."

   (a) Examine the results of your calculations on the equivalent resistance of parallel circuits. Is the above statement consistent with your answers?

   (b) Can you devise any exception(s) to the statement?

Investigation

7.12 Series-Parallel Combination Circuits

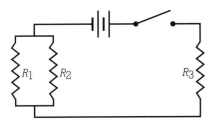

Fig. 17

So far, you have examined circuits in which the resistors were connected either in series or in parallel. Some circuits contain both types of connections. Such circuits are known as **series-parallel combination circuits**. An example is shown in Figure 17.

How is the equivalent resistance of this circuit related to the resistances of the individual resistors? Suggest a mathematical equation and discuss it with your group. Then test your suggestion as follows.

Materials

4 resistors
ammeter
voltmeter
power supply
switch
connecting wires

Method

1. Develop an experimental procedure for determining the equivalent resistance of a series-parallel circuit.

2. Design a series-parallel circuit that includes the materials listed above.

3. Have your teacher check and approve your design.

4. Conduct the experiment and determine the equivalent resistance from the data obtained.

Follow-up

1. Propose a mathematical relationship between the equivalent resistance and the resistances of the individual resistors in your circuit.

2. Use the relationship from question 1 to calculate the equivalent resistance.

3. Compare the value from question 2 with the value you determined experimentally. Account for any differences.

4. What is the resistance of the single resistor that can replace all the resistors in this circuit (Figure 18)?

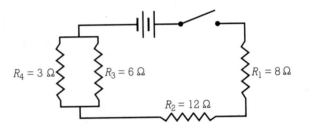

$R_4 = 3\ \Omega$ $R_3 = 6\ \Omega$ $R_1 = 8\ \Omega$

$R_2 = 12\ \Omega$

Fig. 18

5. Calculate the unknown resistance, R_1, in this circuit (Figure 19). The equivalent resistance of the circuit is 5 Ω.

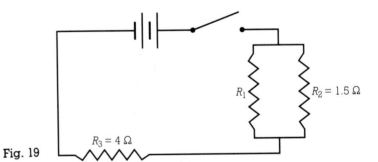

R_1 $R_2 = 1.5\ \Omega$

$R_3 = 4\ \Omega$

Fig. 19

6. Suppose you were given four 6-Ω resistors to connect. At one extreme, you could connect them all in series. At the other extreme, you could connect them all in parallel. There are also intermediate possibilities.
   (a) Calculate the equivalent resistance for each arrangement.
   (b) Which arrangement gives the greatest equivalent resistance?
   (c) Which arrangement gives the smallest equivalent resistance?

7. Write a report on this investigation.

Investigation	# 7.13 Experimental Errors in Resistance Measurements

In previous investigations, you have observed some discrepancies between measured and calculated resistance values. How can we analyze these discrepancies to decide their significance?

Two concepts are very important. They are the **absolute error** and the **relative error** in a measured quantity. Both are measures of the accuracy of experimental results. The absolute error is the difference between the measured and accepted values for a quantity. It is defined as:

Absolute error = measured value − accepted value

So, when the measured value is greater than the accepted value, the absolute error is positive. A negative absolute error arises when the measured value is smaller than the accepted value.

The relative error, often called the **percentage error**, is defined as follows:

$$\text{Percentage error} = \frac{\text{absolute error}}{\text{accepted value}} \times 100\%$$

Suppose, for example, you determine the mass of a standard 1-g mass on a beam balance and obtain a value of 1.05 g. The absolute error in the measurement is 0.05 g.

$$\begin{aligned}\text{The percentage or relative error} &= \frac{0.05 \text{ g}}{1.00 \text{ g}} \times 100\% \\ &= 5\%\end{aligned}$$

If your measured value is 0.95 g, then the absolute error is −0.05 g and the relative error is −5%.

You will now take a closer look at the errors in values you measured earlier.

Materials

power supply
switch
resistors
ammeter
voltmeter
ohmmeter

Method

1. Calculate the absolute error in each equivalent resistance you determined experimentally in investigation 7.10. Assume that the accepted value in each case is the value you calculated from your own mathematical equation. (If you completed investigation 7.12, also calculate the absolute error in the equivalent resistance you determined there.)

2. Use the absolute errors from step 1 to calculate the relative error in each case.

3. Explain how the errors calculated in steps 1 and 2 arose.

4. Take several resistors and determine the resistance of each experimentally, using an ammeter and a voltmeter. Repeat each determination several times.

5. Determine the resistance of each of the same resistors with an ohmmeter. Repeat each determination several times.

Follow-up

1. Ask your teacher to tell you the resistance of the each resistor you used in the investigation (unless the resistors are already colour-coded).

2. Determine the absolute and relative errors in each resistance you determined with the ammeter and voltmeter.

3. Determine the absolute and relative errors in each resistance you determined with the ohmmeter.

4. Which method of determining the resistance was more accurate in each case?

5. Suggest possible causes of errors in the measured resistance values.

6. Suggest ways of improving your measuring technique in order to minimize errors. (If time permits, test your suggestions to see if they work.)

7. In steps 4 and 5 of the method, you repeated each determination several times. Use your data to compare the precision of the procedures in these two steps.

8. Write a report on the investigation.

Electrically Speaking

NON-OHMIC CONDUCTORS

In several experiments, you have assumed that the ratio of potential difference to current intensity for a resistor is the same for all points on a graph of these two variables. In other words, a plot of V against I is a straight line. This is true under normal conditions for resistors made of such materials as copper, aluminum, silver, platinum, and iron, all of which obey Ohm's law. However, some other resistors, such as those that contain liquids and gases, have resistances that change as the potential difference changes. So the ratio $\dfrac{V}{I}$ is not the same for all points on the graph, which is not a straight line (Figure 20). The behaviour of such materials is described as being **non-ohmic**.

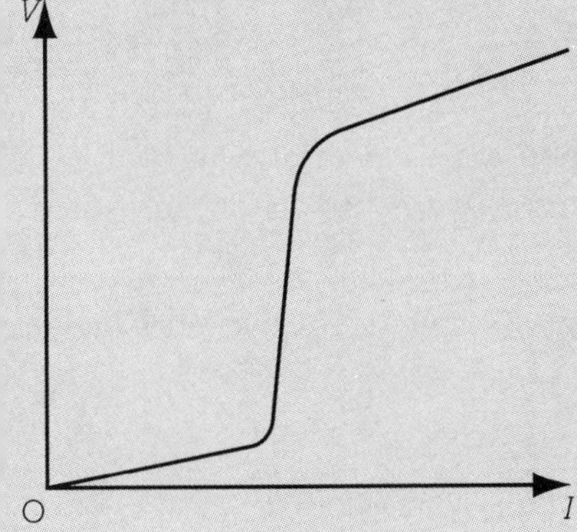

Fig. 20 *Plots of V against I for gases, like neon and argon, have this general shape.*

7.14 Effects of Measuring Instruments on Circuits

You have already observed that measuring instruments have their limitations, in terms of both accuracy and precision. It is often difficult to say how inaccuracies come about. Of course, instruments can be used wrongly. However, there is another important possibility. Perhaps the measuring instrument itself somehow distorts the value of the variable it is measuring. (Such a situation is not unusual. Suppose, for example, that you are riding a bicycle. It has a speedometer that works by contact with a moving wheel. The speedometer itself actually slows the bicycle very slightly, so it distorts the quantity it is designed to measure.) Do electrical instruments also distort the quantities they measure?

Materials

Devise your own materials list as you develop a procedure. Choose from among the measuring instruments you have already used.

Method

1. Develop a procedure to measure a variable in a circuit.

2. Devise a way to estimate the error caused by the measuring instrument itself.

3. Suggest ways of improving the measuring instrument and your measuring technique.

4. Test your ideas experimentally.

Follow-up

1. Give some examples from your own experiences of how measuring techniques or measuring instruments distort measured quantities.

2. Write a report explaining how measurements affect results in scientific research.

| *Investigation* | *7.15 Distribution of Electric Current in Circuits* |

You have measured current intensities in a number of circuits. However, you have not examined how the current is distributed among the various parts of a circuit. Is the current intensity the same in all parts of a circuit, or is the current divided up in some way? If it is divided, in what proportions?

Consider a series circuit containing at least two resistors and decide how you think the current intensity is distributed. Then consider a circuit with at least two resistors in parallel and again suggest what the distribution of the current might be. Discuss your hypotheses with members of your group. You will now test your hypotheses.

Materials

power supply
switch
resistors
connecting wires
ammeter

Method

1. Develop your own procedures for testing your hypotheses.

2. Draw your own circuit diagrams for the circuits you intend to build.

3. Have your teacher check your circuit diagrams before you proceed.

4. Record all your current intensity data.

Follow-up

1. On the basis of your measurements, propose a law regarding the distribution of current in a series circuit.

2. Similarly, propose a law that describes the distribution of current in a parallel circuit.

3. Compare your laws with those of other students. Discuss any discrepancies.

4. A German physicist, Gustav Robert Kirchhoff, suggested that the current intensity is the same throughout a series circuit, but not a parallel circuit. He thought that, for resistors in parallel, the sum of their current intensities equalled the current through the rest of the circuit. On the basis of your data, do you think that Kirchhoff was right or wrong? Explain your reasoning.

5. Write a mathematical equation to relate the current intensities in individual resistors (I_1, I_2, and so on) to the total current, I, in the circuit for:
   (a) resistors in series
   (b) resistors in parallel

6. Write a report on the investigation.

Extension

Test the laws that you devised by constructing a series-parallel circuit and measuring the current intensity distribution.

| *Investigation* | ## 7.16 *Distribution of Electric Potential in Circuits* |

In the preceding section, you determined how the current intensity is distributed in series and parallel circuits. You will now extend your study to include the distribution of electric potential.

Decide how you think the potential difference across two resistors is distributed if they are in series. Similarly, decide what you think the distribution is if the resistors are in parallel. Discuss your ideas with your group. Then test your hypotheses.

Materials

power supply
switch
resistors
connecting wires
voltmeter

Method

1. Develop your own procedures for testing your hypotheses.

2. Draw your own circuit diagrams for the circuits you intend to build.

3. Have your teacher check your circuit diagrams before you proceed.

4. Record all your potential difference data.

Follow-up

1. On the basis of your measurements, propose a law regarding the distribution of potential difference in series circuit.

2. Similarly, propose a law that describes the distribution of potential difference in a parallel circuit.

3. Compare your laws with those of other students. Discuss any discrepancies.

4. Kirchhoff suggested that the potential difference is the same throughout a parallel circuit, but not a series circuit. He thought that, for resistors in series, the sum of the potential differences across them equalled the potential difference across the power supply. On the basis of your data, do you think that Kirchhoff was right or wrong? Explain your reasoning.

5. Write as a mathematical equation your relationship between the potential differences across individual resistors (V_1, V_2, and so on) and the potential difference, V, across the power supply for:
   (a) resistors in series
   (b) resistors in parallel

6. Write a report on the investigation.

Extension

Test the laws that you devised by constructing a series-parallel circuit and measuring the potential difference distribution.

7.17 Calculations on Series and Parallel Circuits

With the laws that you devised in investigation 7.15, you can determine the current distribution in a variety of circuits.

Example 1

A Christmas tree is lit by 12 lamps connected in series to a 120-V source. Each lamp has a resistance of 8.0 Ω. What is the current through each lamp?

Solution

Total resistance, R = 12 $\times$ 8.0 Ω

$$= 96 \ \Omega$$

Current, $I = \dfrac{V}{R}$

$$= \frac{120 \text{ V}}{96 \ \Omega}$$

$$= 1.3 \text{ A}$$

As the electric current is the same throughout a series circuit, the current intensity in each lamp is 1.3 A.

Example 2

Five 110-Ω lamps are connected in parallel to a 110-V household circuit.

(a) What is the equivalent resistance of the lamps?
(b) What is the total current flowing through the circuit?
(c) What is the current flowing through each lamp?

Solution

(a) $\dfrac{1}{R} = \dfrac{1}{110} + \dfrac{1}{110} + \dfrac{1}{110} + \dfrac{1}{110} + \dfrac{1}{110}$

$= \dfrac{5}{110}$

$R = \dfrac{110}{5}$

$= 22\ \Omega$

(b) $I = \dfrac{V}{R}$

$= \dfrac{110\ V}{22\ \Omega}$

$= 5.0\ A$

(c) Since there are five lamps with equal resistance, then the current divides itself into five equal parts. The current through each lamp is 1.0 A.

Note that the answers to both examples are consistent with the laws from investigation 7.16 that describe the distribution of potential difference. In Example 1, the potential difference across each lamp in the series circuit is 1.3 Ω × 8.0 Ω = 10 V (to two significant figures). The potential differences across all 12 lamps sum to 120 V. This value is the potential difference of the source, as expected for a series circuit.

In Example 2, the potential difference across each lamp, 110 V, is the same as the potential difference of the source, as expected for a parallel circuit.

You can use your laws on potential difference distribution to solve other problems.

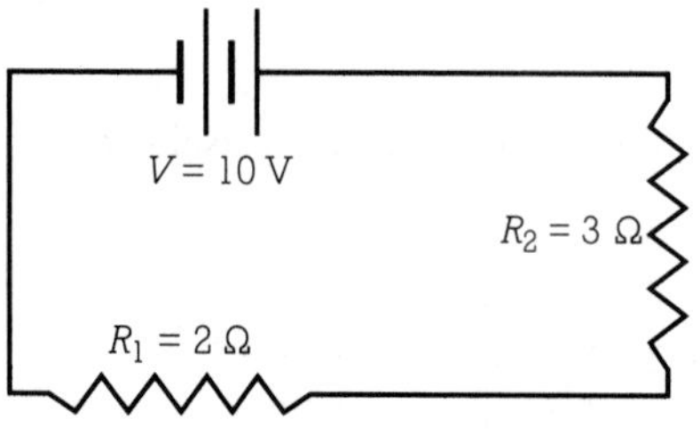

Example 3

Use the circuit diagram (Figure 21) to calculate:

(a) the equivalent resistance
(b) the current, I, through the circuit
(c) the potential differences, V_1 and V_2, across resistors R_1 and R_2, respectively
(d) the difference between the potential difference of the source and sum of V_1 and V_2

Fig. 21

$V = 10\ V$

$R_2 = 3\ \Omega$

$R_1 = 2\ \Omega$

Solution

(a) For resistors in series, equivalent resistance, $R = R_1 + R_2$
$$= 2\,\Omega + 3\,\Omega$$
$$= 5\,\Omega$$

(b) $I = \dfrac{V}{R}$

$$= \dfrac{10\text{ V}}{5\,\Omega}$$

$$= 2\text{ A}$$

(c) $V_1 = IR_1$

$$= 2\text{ A} \times 2\,\Omega$$
$$= 4\text{ V}$$

$V_2 = IR_2$

$$= 2\text{ A} \times 3\,\Omega$$
$$= 6\text{ V}$$

(d) potential difference of source $= 10$ V

$$V_1 + V_2 = 4\text{ V} + 6\text{ V}$$
$$= 10\text{ V}$$

So the difference is zero.

(Remember that, for a series circuit, the sum of the potential differences across the resistors equals the potential difference of the source.)

Example 4

Three resistors, of resistance 6 Ω, 3 Ω, and 2 Ω, are connected in parallel across a 9-V battery. Calculate:
(a) the equivalent resistance
(b) the current through the circuit
(c) the potential difference across each resistor
(d) the difference between the potential difference across each resistor and the potential difference of the battery

Solution

Draw a diagram (Figure 22) to represent the information given.

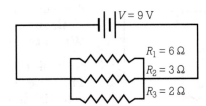

Fig. 22

(a) For a parallel circuit, the equivalent resistance, R, is given by $\dfrac{1}{R} = \dfrac{1}{R_1} + \dfrac{1}{R_2} + \dfrac{1}{R_3}$

$$= \frac{1}{6} + \frac{1}{3} + \frac{1}{2}$$

$$= \frac{6}{6}$$

$$= 1$$

So $\quad R = 1\ \Omega$

(b) $I = \dfrac{V}{R}$

$$= \frac{9\ \text{V}}{1\ \Omega}$$

$$= 9\ \text{A}$$

(c) Remember that the potential difference across each resistor in parallel is the same. It is easiest to determine it from the total current in the circuit and the equivalent resistance. An alternative is to work out the current in one resistor and multiply it by the resistance of that resistor. But this process is difficult and much more time consuming. (Try it!)

$V = IR$
$\quad = 9\ \text{A} \times 1\ \Omega$
$\quad = 9\ \text{V}$

(d) The difference is zero (as expected for a parallel circuit).

Note that it is always a good idea to check that your solution to a problem obeys the laws of current *and* potential difference distribution for the type of circuit you are considering.

As explained in section 7.8, it is also sensible to check the units of all your answers. Rough estimates can be helpful checks in calculations that involve awkward numbers and require the use of a calculator.

Fig. 23

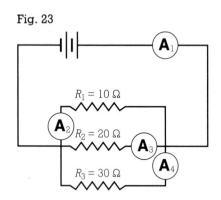

Fig. 24

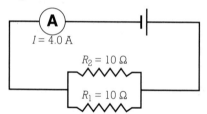

Questions

1. A circuit consists of a battery, three parallel resistors, of 10-Ω, 20-Ω, and 30-Ω resistance, and four ammeters (Figure 23). Which ammeter indicates the smallest current intensity?

2. A circuit (Figure 24) contains two parallel resistors of 10 Ω each. The current through the ammeter is 4.0 A. What is the potential difference across the cell?

Fig. 25

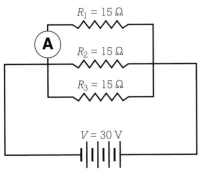

3. What current would flow through the ammeter in this circuit (Figure 25)? What is the total current in the circuit?

4. Three resistors of 30 Ω, 50 Ω, and 40 Ω are connected in series across a 60-V power supply. Calculate:
   (a) the equivalent resistance
   (b) the current in the circuit
   (c) the potential difference across each resistor
   (d) the total potential difference across the circuit

5. Four resistors of 12 Ω each are connected in series to a power supply. The current flowing in the circuit is 5 A. Calculate:
   (a) the equivalent resistance of the circuit
   (b) the potential difference of the power supply
   (c) the potential difference across each resistor

6. Two 20-Ω resistors are connected in parallel to a 40-V power supply. Calculate:
   (a) the equivalent resistance of the circuit
   (b) the total electric current in the circuit
   (c) the electric current through each resistor
   (d) the potential difference across each resistor

Fig. 26

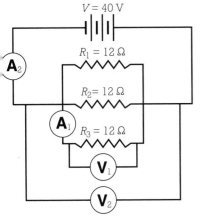

7. Determine the reading on each ammeter and voltmeter in Figure 26.

7.18 Calculations on Series-Parallel Circuits

If you have completed the extension activities in investigations 7.15 and 7.16, you are in a position to solve numerical problems that involve series-parallel circuits.

Example 1

Fig. 27

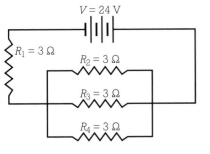

Four 3-Ω resistors are connected to a 24-V battery. Three of the resistors are connected in parallel. The fourth is in series with them (Figure 27).

Determine:
(a) the equivalent resistance of the three resistors in parallel
(b) the total resistance of the circuit
(c) the total current in the circuit
(d) the current intensity in each resistor

Solution

(a) The equivalent resistance of the parallel resistors, R', is given by

$$\frac{1}{R'} = \frac{1}{3} + \frac{1}{3} + \frac{1}{3}$$
$$R' = 1\ \Omega$$

(b) The total resistance $= 1\ \Omega + 3\ \Omega$
$$= 4\ \Omega$$

(c) $I = \dfrac{V}{R}$

$$= \frac{24\ \text{V}}{4\ \Omega}$$
$$= 6\ \text{A}$$

(d) The resistor in series experiences a current intensity of 6 A. The current splits equally between the three resistors in parallel, so the current intensity in each is 2 A.

Example 2

Examine the series-parallel circuit shown (Figure 28). Determine the voltmeter reading in the circuit.

Solution

The equivalent resistance, R', of the parallel resistors is given by

$$\frac{1}{R'} = \frac{1}{3.0} + \frac{1}{3.0}$$
$$= \frac{2}{3.0}$$
$$R' = 1.5\ \Omega$$

So the equivalent resistance of the circuit is $R = 3.0\ \Omega + 1.5\ \Omega$
$$= 4.5\ \Omega$$

The current in the circuit, $I = \dfrac{V}{R}$

$$= \frac{9.0\ \text{V}}{4.5\ \Omega}$$
$$= 2.0\ \text{A}$$

The potential difference, V_1, across the parallel resistors is

Fig. 28

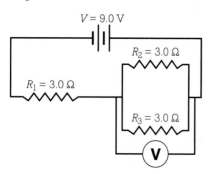

$V = 9.0\ \text{V}$

$R_2 = 3.0\ \Omega$

$R_1 = 3.0\ \Omega$

$R_3 = 3.0\ \Omega$

V

given by $V_1 = IR_1$
$$= 2.0 \text{ A} \times 1.5 \text{ }\Omega$$
$$= 3.0 \text{ V}$$

So the voltmeter reading would be 3.0 V.

Questions

1. Two 50-Ω resistors, R_1 and R_2, are connected in parallel. This parallel set up is connected in series with a 15-Ω resistor, R_3. The circuit has a 120-V source. Calculate:
   (a) the equivalent resistance of R_1 and R_2
   (b) the total resistance of the circuit
   (c) the total current in the circuit
   (d) the potential difference across the 15-Ω resistor
   (e) the potential difference across the two parallel resistors
   (f) the current flowing through each of the parallel resistors

2. The circuit diagram shows a 9.0-V battery and three 3.0-Ω resistors (Figure 29). What is the voltmeter reading when the switch is closed?

3. A 15-Ω heater and 30-Ω frying pan are connected in parallel. The combination is in series with a 20-Ω toaster on a 120-V circuit. Calculate:
   (a) the equivalent resistance of the circuit
   (b) the total current in the circuit
   (c) the potential difference across the parallel combination
   (d) the current in the heater

4. A 40-Ω electric hair dryer is in parallel with a 10-Ω TV set. This combination is in series with a 12-Ω light bulb. If the circuit is supplied with 120 V, calculate:
   (a) the equivalent resistance of the circuit
   (b) the total current in the circuit
   (c) the potential difference across the parallel combination
   (d) the current in the TV set

Fig. 29

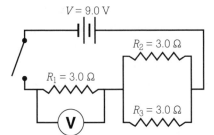

$V = 9.0 \text{ V}$

$R_2 = 3.0 \text{ }\Omega$

$R_1 = 3.0 \text{ }\Omega$

$R_3 = 3.0 \text{ }\Omega$

V

7.19 A Model of Potential Difference

When we deal with abstract concepts, we often develop models to help our understanding. For example, we can get a simplified idea of the insides of an atom by using the solar system as a model. The concept of potential difference is quite complex. What is your present concept of what potential difference means? Discuss your ideas with your group. Are there models we can use to understand potential difference better?

CHAPTER 7

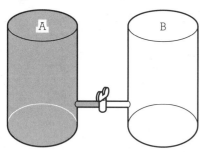

Fig. 30

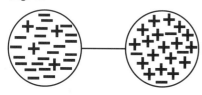

Fig. 31

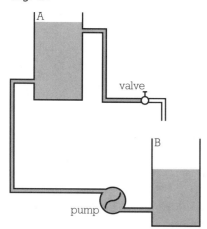

Fig. 32

Questions

1. In Figure 30, two containers, A and B, are connected by a pipe and a valve. Container A is filled with water. Describe what happens when the valve is opened. What causes it to happen?

2. Consider two spheres charged so that one is negative (has an excess of electrons) and the other is positive (has a deficit of electrons). If you connect the spheres with a conducting wire (Figure 31):
   (a) In which direction do electrons flow?
   (b) What causes the electrons to flow?
   (c) When do electrons cease to flow?
   (d) How does the concept of water flow in question 1 compare with the concept of electron flow in (a) to (c)?

3. Suppose that two tanks of water, A and B, are placed so that A stands at a higher level than B (Figure 32). A pipe with a pump leads from the bottom of B to the bottom of A. A pipe with a valve leads from A to B. What happens when the pump is started? What happens when the valve is opened?

4. If a battery is connected in series with a resistor (light bulb) and a switch, what happens when the switch is closed? Compare your answer to your answers in question 3. In what ways are the situations similar? What causes the current to flow?

5. As you have seen, the water pressure model is a useful analogy for simple circuits. But can we extend it to more complex situations? In particular, can we represent the distribution of potential difference in a circuit? Consider Figure 33, which shows a tank of water connected to a water pipe. A series of identical gauges, G_1 - G_4, is attached to the pipe. The gauges all register the same pressure if the valve, V, is closed.

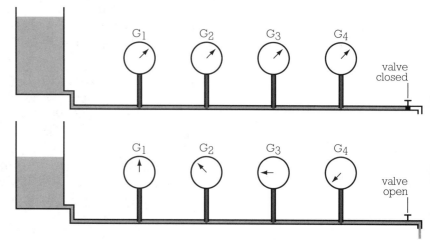

Fig. 33

Fig. 34

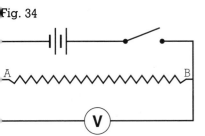

When V is opened, water flows at the same rate past every point in the pipe. However, the gauges indicate a drop in pressure along the pipe as water travels further from the tank. Compare this situation with a series circuit containing a high resistance wire, AB, a battery, and a switch. A voltmeter is connected in parallel across the ends of AB (Figure 34).

The voltmeter reading indicates the potential difference of the battery, which equals the potential difference across the high resistance wire. But what if one contact of the voltmeter is gradually moved from B toward the other contact at A? The voltmeter reading will gradually drop. (Remember that the potential difference across a series circuit is the sum of the potential differences across its parts.) In what ways are the situations in the pipe and the series circuit similar?

6. You have seen some similarities between the concepts of potential difference and water pressure. But no model is a perfect resemblance of what it represents. In what ways is water pressure different from potential difference?

Investigation

7.20 Behaviour of Unknown Circuits

Suppose that you are faced with the task of modifying or duplicating a circuit, but you have no circuit diagram available to you. The connecting wires are hidden from view, perhaps in the walls of a building (Figure 35). How can you go about constructing a circuit diagram for the unknown circuit?

The following investigation begins with an examination of some important properties of series and parallel circuits. These properties will help you solve a "mystery circuit" presented by your teacher.

Materials

power supply
two flashlight bulbs
switch
mystery circuit
materials for building a mystery circuit
ammeter
voltmeter

Method

Fig. 35 *Electrical wiring is often hidden from view.*

1. Connect the power supply, bulbs, and switch, with the two bulbs in series. Close the switch.

2. Open the switch, unscrew one of the bulbs, and close the switch again. What do you observe?

3. Repeat steps 1 and 2, but this time connect the two bulbs in parallel.

4. Your teacher will now present you with a mystery circuit. You will be able to see bulbs and switches, but not the wires that connect them. Open and close switches, and remove and replace bulbs to find out what the connections are. You will need to be very well organized to keep track of the results. Use a sketch and/or a chart to keep track.

5. If necessary, take measurements with the ammeter and voltmeter, and use your knowledge of Ohm's law to aid your deductions.

6. Draw a circuit diagram for the mystery circuit.

7. Build the mystery circuit and check that it behaves in the same way as the circuit you were given. If it does not, then modify your circuit diagram and circuit as necessary.

8. Design and build your own mystery circuit for other students to solve.

9. Solve several mystery circuits designed by other students.

Follow-up

1. An electrical technician must replace a burned out 7-Ω resistor. However, only 2-Ω and 4-Ω resistors are available. Which of the following combinations could the technician use to provide a resistance of 7 Ω?

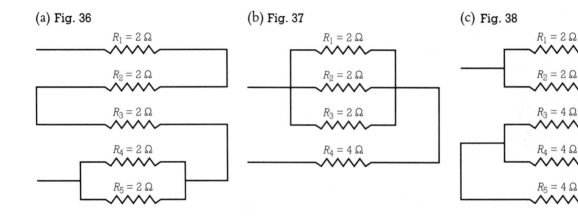

(a) **Fig. 36**

$R_1 = 2\ \Omega$
$R_2 = 2\ \Omega$
$R_3 = 2\ \Omega$
$R_4 = 2\ \Omega$
$R_5 = 2\ \Omega$

(b) **Fig. 37**

$R_1 = 2\ \Omega$
$R_2 = 2\ \Omega$
$R_3 = 2\ \Omega$
$R_4 = 4\ \Omega$

(c) **Fig. 38**

$R_1 = 2\ \Omega$
$R_2 = 2\ \Omega$
$R_3 = 4\ \Omega$
$R_4 = 4\ \Omega$
$R_5 = 4\ \Omega$

2. Construct the assembly you chose in question 1 and measure its equivalent resistance.

3. Compare your measured value in question 2 with the calculated value of 7 Ω.

4. Explain any difference between the calculated and measured values.

P O I N T S · T O · R E C A L L

- An electric current moves easily through a conductor but not through an insulator.
- A circuit is a path along which an electric current can flow.
- The conductance of an object depends on several factors.
- Resistance is a measure of the difficulty an electric current experiences as it flows through an object.
- The potential difference of a source pushes an electric current through a circuit.
- The SI unit of electric current is the ampere (A).
- The SI unit of potential difference is the volt (V).
- The SI unit of resistance is the ohm (Ω).
- The SI unit of conductance is the siemens (S).
- For any object, the conductance in siemens is the reciprocal of the resistance in ohms.
- Materials with zero resistance at low temperatures are called superconductors.
- An ammeter has a low resistance and is connected in series with the device whose current intensity it measures.
- A voltmeter has a high resistance and is connected in parallel with the device whose potential difference it measures.

- Ohm's law states that, for a given resistor at constant temperature, the ratio of potential difference to electric current is a constant. This constant is the resistance of the resistor.
- All devices in series receive the same electric current.
- All devices in parallel experience the same potential difference.
- For devices in series, the equivalent resistance is the sum of individual resistances.
- For devices in parallel, the reciprocal of the equivalent resistance equals the sum of the reciprocals of the individual resistances.
- For resistors in series, the equivalent resistance is greater than the resistance of any individual resistor.
- For resistors in parallel, the equivalent resistance is smaller than the resistance of any individual resistor.
- Errors in data can arise in many ways, including the distorting effect of a measuring instrument on the variable it is measuring.
- Flowing water and water pressure can be useful analogies for electric current and potential difference.

R E V I E W · Q U E S T I O N S

1. What factors determine the conductance of a wire at constant temperature?
2. What happens to the conductance of a toaster as it warms up? Explain your answer.
3. Why are electrical wires generally made of copper, rather than iron?
4. State Ohm's law.
5. Use Ohm's law to define the following units:
   (a) ampere   (b) volt   (c) ohm
6. Are car headlights connected in series or in parallel? Explain how you know the answer.
7. Draw a circuit diagram showing two light bulbs, a switch, and a power supply. Connect them in such a way that you can operate both light bulbs at the same time.
8. Draw a circuit diagram showing two light bulbs, two switches, and a power supply. Connect them in such a way that you can operate the light bulbs independently.
9. An ammeter is connected in series with two 100-Ω light bulbs that are in parallel. When two additional bulbs are connected in parallel with the others, how does this change affect:
   (a) the reading on the ammeter?
   (b) the resistance of the circuit?
10. Calculate the current through a toaster having 11-Ω resistance and connected to a 110-V circuit.
11. What is the conductance of a circuit that has a resistance of 10 Ω?
12. What is the resistance of a conductor that carries 3.0 A of current when connected to a 6.0-V battery?
13. What is the resistance of an electric oven that draws 20 A from a 220-V line?
14. What is the equivalent resistance of 4.0-Ω, 3.0-Ω, and 12-0 Ω devices if you connect them:
    (a) in series?   (b) in parallel?
15. Graph the following data and determine:
    (a) conductance   (b) resistance

Potential difference (volts)	Electric current (amperes)
2.4	1.0
3.7	1.5
6.1	2.5
8.7	3.5

16. Suppose you immerse a heating coil in water and adjust the coil to keep the water temperature constant. You obtain the following data for the coil:

Potential difference (volts)	Electric current (amperes)
6.20	0.20
9.36	0.30
10.92	0.35
12.48	0.40

Does the behaviour of the heating coil obey Ohm's law?

17. Explain why an ammeter has a very low resistance and is connected in series, whereas a voltmeter has a very high resistance and is connected in parallel.
18. Eight 360-Ω light bulbs are connected in parallel to a 120-V circuit. Calculate:
    (a) the equivalent resistance of the circuit
    (b) the current intensity in each light bulb
    (c) the total current intensity in the circuit
19. A source with a potential difference of 2.0 V is connected to two resistors, of 6 Ω and 9 Ω, in parallel. In series with the source is a 2-Ω resistor. Calculate the current intensity in the 6-Ω resistor.

20. A current of 4.0 A is flowing through a 2-Ω resistor, which is in series with a parallel combination of a 4-Ω and a 6-Ω resistor. Calculate:
    (a) the current intensity in each of the parallel resistors
    (b) the potential difference across the entire circuit
21. A circuit contains a 10-Ω resistor in series with a parallel arrangement of three 30-Ω resistors (Figure 39). The ammeter reads 2.0 A when the switch to one of the parallel resistors is open. What will the ammeter reading be when the switch is closed?

Fig. 39

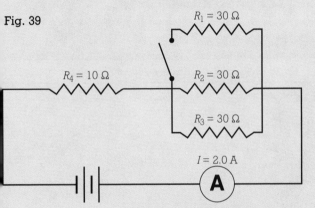

22. What factors affect the resistance of a wire at constant temperature, and in what way does the resistance depend on each factor?
23. A 24-V lamp has a resistance of 8.0 Ω. What resistance must be placed in series with it if it is to be plugged into a 120-V supply?
24. Lamps 1, 2, and 3 have the same resistance. They are connected in a series-parallel combination to a power supply, with lamp 1 in series with the parallel arrangement of 2 and 3.
    (a) Draw a circuit diagram for this arrangement.
    (b) Describe the relative brightness of the three lamps. Explain your reasoning.
    (c) If lamp 3 burns out, what change occurs in the relative brightness of the two remaining lamps. Explain your reasoning.

25. Determine the unknown values on the circuit diagram (Figure 40).

Fig. 40

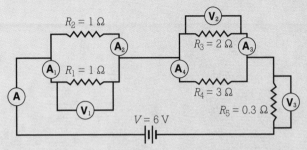

26. In order for a fuse to protect a circuit properly, should you connect it to the circuit in series or in parallel? Explain your answer.

CHAPTER

8

ELECTRICAL ENERGY: PRODUCTION, TRANSFORMATION, AND TRANSMISSION

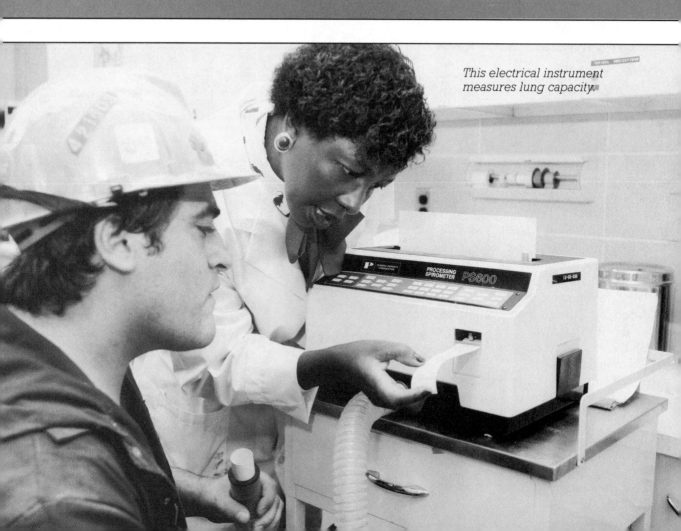

This electrical instrument measures lung capacity.

CONTENTS

8.1 Electrical Energy

We usually take our energy supply for granted. However, it does not take much thought for us to realize how dependent we are on various forms of energy, electrical energy in particular. Try the following activity and construct a chart to show your findings:

1. List the devices you use every morning between the time you get up and the time you get to school.

2. Indicate the type(s) of energy used by each device.

3. Describe the energy transformation(s) that occur in each device.

4. Indicate the devices that do not require the use of electricity.

5. Discuss your findings with others in your class.

You should now be aware that electricity is used so widely that it is difficult to imagine life without it. We buy it for the work it can do for us in various ways, but we usually do not think about where the electricity comes from. We usually take for granted the devices that convert it into heat, light, sound, mechanical energy, chemical energy, and energy of motion.

Perhaps one reason that we do not notice our electrical energy supply is that it is so easy to use. A switch turns it on and off, and most electrical devices take little effort to operate. In this chapter, we will take a closer look at our electrical energy supply. In particular, we will examine how and where electrical energy is produced, how it is delivered to us, and how much electrical energy we use.

8.2 Energy and Power

In any discussion of energy use, it is important to distinguish the meanings of two terms. The first term is **energy** itself. This is defined as the ability to do work. For example, an electric current, which supplies electrical energy, does work when it runs a motor or heats a wire. The devices convert the energy of moving electrons into other forms, like a water wheel converts the energy of moving water into other forms.

Power, on the other hand, is defined as the rate of using energy or the rate of doing work. If we compare two electrical machines, the more powerful one uses electrical energy faster than the less powerful one. For example, a toaster uses more energy than a flashlight over the same period of time.

Since power is the rate of using energy, we can represent it by an equation:

$$\text{Power} = \frac{\text{energy}}{\text{time}} \quad \text{or} \quad P = \frac{E}{t}$$

As you can see from this relationship, the quantity of energy a machine uses depends on the time for which it operates. (Since $E = Pt$, the longer t is, the greater E is.) So it takes more energy to make a darker piece of toast than a lighter piece of toast in the same toaster. It is possible for a less powerful machine to use more energy than a more powerful one, but only if the less powerful machine is on longer.

Like other physical quantities, energy and power are measured in standard units. The SI unit of energy is the joule, symbol J. One joule is a very small quantity of energy. A raw carrot gives you about 200 000 J of energy when you eat it. Even when you are

$1\,kJ = 1000\,J$
$1\,MJ = 1\,000\,000\,J$

asleep, your body uses about 70 J of energy every second to keep going! Because the joule is so small, it is common to measure quantities of energy in kilojoules, kJ, or megajoules, MJ.

The SI unit of power is the watt, symbol W. A watt is equivalent to one joule per second. So, another way of saying that your body uses seventy joules of energy per second (70 J/s) while you are sleeping is to say that your body uses 70 W of power. By contrast, an electric toothbrush has a power rating of about 3 W, meaning that it uses 3 W of power.

Since the joule is a small unit of energy, the watt is a small unit of power. Therefore, kilowatts (kW) and megawatts (MW) are commonly used. A typical electric kettle has a power rating of 1200 W, or 1.2 kW. The Manic 5 electrical generating station in Northern Québec has a maximum power output of 1300 MW.

A light bulb is perhaps the commonest type of electrical device whose power we routinely describe. You are all familiar with the different ''power ratings'' of light bulbs. The commonest ones in the home are rated at 40 W, 60 W, and 100 W. They use 40 J, 60 J, and 100 J of electrical energy per second, respectively.

For electrical appliances, you can check the power rating on the **rating plate**, often found on the back of or underneath the appliance (Figure 1).

For example, the rating plate on a typical toaster may show the information ''120 V, 840 W.'' Clearly, the toaster, designed for a 120-V supply, has a power rating of 840 W.

Panasonic
MODEL NO. RX - 5011
POWER SOURCE
AC ∿ 120V 60Hz 16W
BATTERY 6 " D " SIZE BATTERIES   9V
(**Panasonic** UM- 1 OR EQUIVALENT)

Matsushita Electric Industrial Co., Ltd.
Made in Japan
Part No. T 886ZA

Fig. 1

ELECTRICAL TIDBITS

ALTERNATING CURRENT

You will find ''60 Hz'' written on many rating plates. This reflects the fact that the current supplied to our homes is **alternating current**, that is, current that regularly reverses its direction. (By contrast, the **direct current** you draw from a battery does not change direction.) The figure ''60'' indicates the number of times the generator coil rotates each second. As there are two reversals of current direction for each rotation of the coil, a 60-Hz electric current has 120 reversals of current direction per second. In a light bulb, the light intensity changes as the current goes through its repeating cycles. However, we do not see a flickering effect because the process is too fast for our eyes to detect.

Questions

1. Examine the rating plate on an electrical appliance at school or in your home. What information is recorded on the plate?

2. How much energy, in joules, does the appliance use each second?

3. What is the power rating of the appliance in kilowatts?

8.3 Electrical Energy and Power Rating

No doubt you have seen energy meters on the outsides of buildings. They record how much energy is used in a certain period of time (Figure 2).

Fig. 2

The standard unit of energy is the joule. However, meters like the one in Figure 2 use another unit of energy, namely, the kilowatt hour. Its standard symbol is kW·h, though the dot (which shows that two units are being multiplied) is sometimes omitted to give kWh.

As you learned in section 8.2, the quantity of electrical energy used by an appliance depends on both its power rating and the length of time it is turned on. (Remember $E = Pt$.) It is obvious that a quantity of energy in kilowatt hours is obtained by multiplying a power rating in kilowatts by a time measured in hours. For example, suppose you use a 1100-W electric iron for two hours. The energy consumed is given by the relationship $E = Pt$.

$$\begin{aligned} \text{So} \quad E &= 1100\,\text{W} \times 2.0\,\text{h} \\ &= 2200\,\text{Wh} \\ &= 2.2\,\text{kWh} \end{aligned}$$

Of course, we could first convert the power rating to kW, giving:

$$\begin{aligned} 1100\,\text{W} &= 1.1\,\text{kW} \\ \text{So} \quad E &= 1.1\,\text{kW} \times 2.0\,\text{h} \\ &= 2.2\,\text{kWh} \end{aligned}$$

It is not obvious how to read energy consumption from an energy meter. The four numbered dials are read from left to right, but the first and third dials move counterclockwise, whereas the other two move clockwise. A final complication is the fact that, once the numbers from the four dials are recorded, the result is multiplied by 10 to give the energy reading in kilowatt hours. For example, in Figure 3, the readings on the four dials, from left to right are 1, 5, 8, and 2. The overall reading is therefore 1582, which we multiply by 10 to give a final value of 15 820 kWh. The quantity of energy used in a given time period equals the value determined at the end of the period minus the value determined at the beginning.

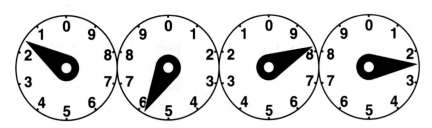

Fig. 3

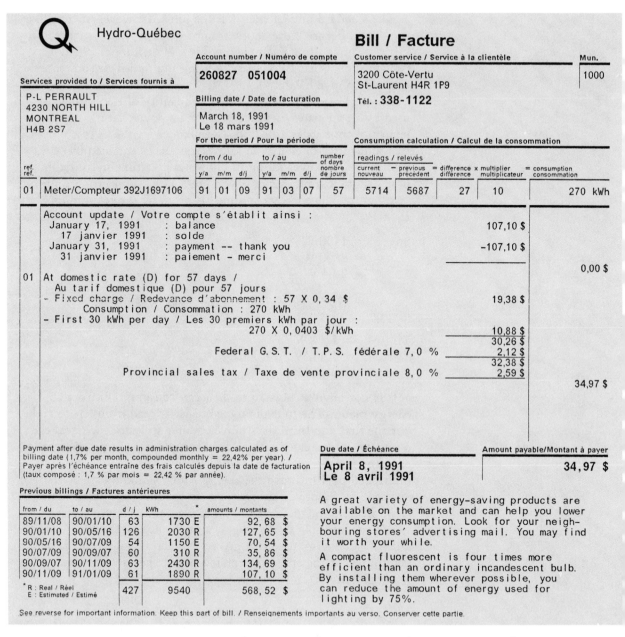

Fig. 4   The invoice a homeowner or business receives from a utility company shows the quantity of energy used in a certain period (Figure 4). Note that this particular utility company uses the formula

current reading – previous reading
= difference × multiplier
= consumption

In other words, instead of taking the meter reading on each date, multiplying it by ten, and then taking the difference, this company takes the difference in the meter readings and multiplies it by ten. Either way, the final answer is the same. (If you are unsure about this, work the consumption out both ways and compare your answers.)

Many utility companies publish a list of average energy consumption values for various appliances. As the figures are averages, they are only a rough guide to the actual energy consumption of an appliance. Different versions of the same type of appliance have different power ratings, and different people use the same type of appliance for different lengths of time. However, we can use the list to work out the average cost of operating appliances. As you can see from the following table, some types of appliances typically use much more energy than others.

Average Energy Consumptions of Appliances		
Appliance	Rating (W)	Average/month (kWh)
(a) *High consumption:*		
air conditioner (window)	1000	275
central air conditioner	2000	480
dryer	5000	80
refrigerator-freezer (0.4 m³)	350	150
water heater, Cascade 60	4500	400
(b) *Low consumption:*		
black-and-white TV	100	12
built-in oven	3000	50
coffee maker	600	9
colour TV	300	36
dishwasher	1500	30
fluorescent light (2 × 40 W)	80	12
incandescent light	100	15
microwave oven	1200	30
washing machine	500	10
(c) *Very low consumption:*		
hair dryer	1000	3
iron	1100	5
radio	60	5
toaster	1200	3

To take just one example from this table, a colour TV uses 36 kWh of electrical energy per month, on average. In a city where electrical energy costs, say, 5 cents per kilowatt hour, the average cost of running the TV for a month is 36 × 5 cents, or $1.80.

Questions

1. The two sets of dials in Figure 5 are schematic diagrams of the readings on an electric meter on two different dates.

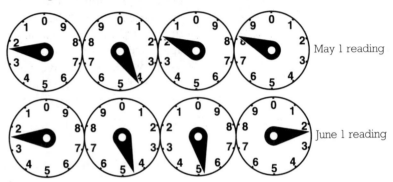

Fig. 5

(a) What is the meter reading on May 1?
(b) What is the meter reading on June 1?
(c) How much energy, in kilowatt hours, was used during this period?
(d) Calculate the cost of the electrical energy used if each kilowatt hour cost 4.03 cents.

2. A colour TV has a power rating of 300 W. What is the cost of watching this TV for two hours, if you are charged 4.03 cents per kilowatt hour?

3. A coin-operated clothes dryer, rated at 4 kW, runs for 15 min when you insert 75 cents.
(a) If the cost of energy is 4.03 cents per kilowatt hour, what is the cost of the energy used by the dryer in one 15-min cycle?
(b) How much profit does the machine make per cycle? (In reality, the profit is less than you think, because of other costs involved.)

4. For each of the following appliances, running for the specified time period, what is:
(a) the quantity of energy used?
(b) the cost at 4.03 cents per kilowatt hour?
(i) a 400-W drill for 2 h
(ii) a 300-W air conditioner for 8 h
(iii) a 750-W block heater for 7 h

5. Use the table of average energy consumptions to calculate the average cost of using a dishwasher for one year. Assume that electrical energy costs 4.03 cents per kilowatt hour.

ENERGUIDE RATINGS

The energy consumption table represents only average values for various types of appliances. Design features, size, and other factors can affect the energy consumption of appliances. Major appliances, such as ovens and refrigerators, now carry "Energuide" stickers that tell you the typical monthly electrical energy consumption of that particular model (Figure 6).

As an outside activity, check some Energuide stickers in stores to find out how much the energy consumption varies from one model to another of the same type of appliance. For the models with the highest and lowest Energuide ratings, compare the annual running costs.

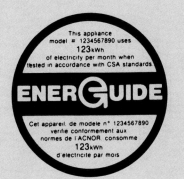

Fig. 6

| Investigation | ## 8.4 A Unit for Electric Current |

In investigation 3.4, you examined ways of producing electric charges and studied the properties of charged objects. But exactly what is an electric charge? Also, how is the concept of electric charge related to the idea of an electric current? Write down your own ideas and discuss them with your group. Now watch the demonstration that your teacher will carry out and answer the follow-up questions.

Materials

electrostatic generator
2 plates mounted on plastic or rubber stands
pithball suspended from a ring stand
connecting wires
glass rod

Method

1. Connect the terminals of your electrostatic generator to the two metal plates (Figure 7).

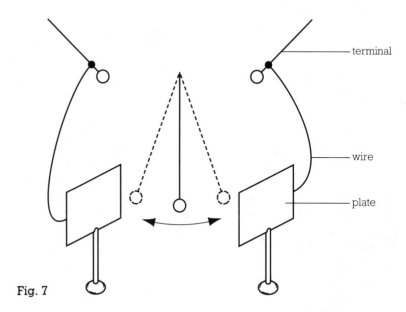

terminal

wire

plate

Fig. 7

2. Place the two plates 10 cm to 15 cm apart.

3. Suspend the pithball between the two plates, so that the ball lines up with the centres of the plates.

4. To charge the plates, crank the generator, with its terminals far away from each other. (If the terminals are too close together, the accumulated charge will cause a spark to jump between them.) The work you do when you crank the generator separates electrons from their atoms.

5. Lift the pithball with a glass rod to bring it close to one of the plates. Then remove the rod.

6. Observe the pithball as the handle is cranked. What is the effect on the pithball of cranking faster?

Follow-up

1. (a) What is transferred to the pendulum when it touches the negative plate?
   (b) What is tranferred from the pendulum when it touches the positive plate?

2. As the generator was cranked faster, the quantity of charge generated per second went up.
   (a) What was the effect on the pendulum?
   (b) Was the quantity of charge transferred from one plate to the other affected? If so, in what way?
   (c) Did the current through the circuit change? If so, in what way?

3. From your answer to question 2, decide whether the current through the circuit depends on:
   (a) the quantity of charge transferred
   (b) the time it takes to transfer a given quantity of charge

4. Suggest an equation that relates the current, I, to the quantity of charge, Q, and the time the current flows, t.

5. We measure electric charge in units called coulombs, symbol C. The ampere, the standard unit of electric current, is defined as a flow rate of one coulomb of electric charge per second. Is this definition consistent with the equation you derived in question 4? Explain your answer.

6. Research Andre Ampère's work. How is it applied to modern technology?

The charge on a single electron is very small, only 1.6×10^{-19} C.

Investigation

8.5 Potential Difference Across Batteries in Series

In section 8.4, you added to your understanding of electric charge and electric current. You can further increase your knowledge of these topics, and broaden your concept of potential difference, by means of the following investigation. It involves connecting batteries in series. Such an arrangement has some common applications. In fact, you connect batteries in series in any flashlight or portable tape player that takes two or more of them.

Before you begin the investigation, suggest answers to the following questions, and discuss the answers with your group:

If you connect several batteries in series, how is the total potential difference across the batteries related to their individual potential differences?

How is the current in an electric circuit of known resistance related to the potential difference across the circuit?

How does a change in potential difference affect the charges moving through a circuit?

Materials

several batteries
voltmeter
ammeter
flashlight bulb
connecting wires
switch

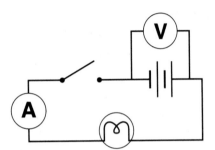

Fig. 8

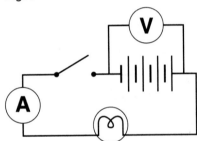

Fig. 9

Method

1. With the voltmeter, measure the potential difference across one battery.

2. Tape the positive terminal of one battery to the negative terminal of another. Measure the total potential difference across the connected batteries.

3. Add a third battery and repeat the potential difference measurement.

4. Connect a battery, light bulb, ammeter, voltmeter, and switch (Figure 8). Close the switch for a few seconds and record the potential difference and current readings. Take note of the brightness of the bulb.

5. Connect a second battery in series with the first (Figure 9) and repeat the measurements.

6. Add a third battery and repeat.

Follow-up

1. What is the mathematical relationship between the potential difference across several batteries in series and their individual potential differences?

2. A police flashlight takes six 3-V batteries. A police officer loads in the batteries but places one battery in backward. Will the flashlight work? Explain your reasoning.

3. What was the effect of increasing the potential difference on
   (a) the brightness of the light bulb?
   (b) the current flowing through the circuit?

4. Research the contribution of Alessandro Volta's invention of the battery to the application of electromagnetic principles in the nineteenth century.

| Investigation | ## 8.6 The Relationship of Potential Difference and Resistance |

Materials

variable power supply
ammeter
voltmeter
several resistors of known resistance
switch
connecting wires

Method

1. Construct a circuit that contains an ammeter, voltmeter, and resistor. Record the potential difference of the source, the current through the circuit, and the value of the resistor.

2. Replace the resistor with another of different resistance and change the potential difference of the source such that the current remains unchanged. Again record your data.

3. Repeat step 2 several more times with different resistors, recording your data each time.

4. Tabulate your data and plot a graph of V against R to show the relationship between the resistance and the potential difference when the current is constant.

Follow-up

1. (a) Measure the slope of the graph.
   (b) Compare the value of the slope to the current.

2. (a) Does your answer to question 1(b) make sense in terms of your earlier knowledge? Explain. (Recall section 7.7.)
   (b) State the mathematical relationship between potential difference, resistance, and current.

3. Use your graph to determine the potential difference when the resistance is 7 Ω.

4. As the resistance increases, the quantity of work (energy) required to push the current around the circuit increases.
   (a) What is responsible for this increased work?
   (b) How is the potential difference related to the quantity of energy used in the circuit?

5. Power can be defined by the relationship $P = VI$, or

power (watts) = potential difference (volts)
  $\times$ electric current (amperes)

Is this relationship consistent with your answer to question 4(b)?

6. Use the equation provided in question 5 to determine the power rating of an appliance that draws 3.0 A of current from a 120-V circuit.

7. If a toaster designed for a 120-V supply has a power rating of 840 W, what is the current intensity through the toaster?

8.7 Energy and Units of Measurement

You will recall from earlier sections that energy is the ability to do work and that the SI unit of energy is the joule. You have also learned that another unit, the kilowatt hour, is commonly used to measure electrical energy.

By now, you have encountered several relationships that apply to electric circuits. For example, you know that energy = power $\times$ time ($E = Pt$). Also, power = potential difference $\times$ electric current ($P = VI$). Now you can use these relationships to propose another unit of electrical energy.

Questions

1. From the two equations in the previous paragraph, derive a single equation that relates electrical energy to potential difference, electric current, and time.

2. From the equation you derived in question 1, can you suggest another unit for electrical energy? Compare your suggestion with those of other students.

3. From previous sections, you know that 1 V = 1 J/C, and 1 A = 1 C/s. Determine how the unit you proposed compares with the joule.

4. You know that 1 W = 1 J/s. Therefore, 1 J = 1 Ws. Knowing this, determine:
   (a) how many joules are equivalent to 1 kWh
   (b) how many megajoules are equivalent to 1 kWh

5. A 110-V percolator uses 8.0 A of current for 8.0 h. Calculate:
   (a) the power rating of the percolator
   (b) the electrical energy it uses in:
   (i) kilowatt hours   (ii) megajoules   (iii) watt seconds

6. Calculate the power rating of each of the following.
   (a) an electric train that draws 0.3 A from a 9.0-V battery
   (b) an electric typewriter that operates at 1.2 A and 110 V

8.8 Transmission of Electrical Energy

A potential difference across a resistor pushes an electric current
through it. You learned in investigation 8.6 that, the greater the
resistance, the greater the potential difference needed to force the
same current through it.

A term sometimes used to describe the potential difference across
one resistor is ''potential difference drop.'' In a long resistor, the
potential difference drop can be very great. Since we send electricity
from one place to another along very long transmission lines
(Figure 10), we have to think about ways of doing this as
efficiently as possible.

Fig. 10

As electrons move along metal wires, they collide with metal ions in the wires. These collisions create heat. So, some of the energy of the electrons is lost in the form of heat. This is known as the "Joule effect." This conversion of electrical energy into heat results in a loss of power in the wires.

You know that $P = IV$, so, for a certain current through a wire, the power loss must depend on the voltage drop across the wire. Recall that Ohm's law tells you the relationship between the voltage drop, the current, and the resistance, $V = IR$. We can use these two relationships to work out how to minimize the power loss in transmission wires.

Questions

1. Use the two equations from the previous paragraph to derive an equation that relates the power loss in a wire to its resistance and the current intensity. (The equation you derive represents Joule's law.)

2. Use the equation from question 1 to explain why:
   (a) electrical transmission lines are thick
   (b) it is best to transmit electricity at a low current intensity

8.9 High-Tension Wires

You have probably heard electrical transmission lines referred to as "high-tension wires." This just means that the potential difference across them is very high. Why should this be the case? What is the advantage of a high potential difference?

You learned in the previous section that less power is lost if the current intensity is small. For a given resistance, the higher the potential difference, the lower the current intensity. Therefore, a higher potential difference gives a lower power loss. But how much does the potential difference affect the power loss? The effect can be quite dramatic, as the following example will show.

Example 1

The transmission lines used to distribute 100 kW of power have a resistance of 0.25 Ω. For a potential difference of:
(a) 500 V   (b) 50 kV
calculate:
 (i) the current intensity in the lines
 (ii) the power loss in the lines
(iii) the percent power loss

Verify the units of the answers in parts (i) and (ii) by means of unit analyses.

Solution

(i) (a) $P = IV$

$$I = \frac{P}{V}$$

$$= \frac{1 \times 10^5 \text{ W}}{5 \times 10^2 \text{ V}}$$

$$= 200 \frac{\text{W}}{\text{V}}$$

$$= 200 \text{ A}$$

(b) $P = IV$

$$I = \frac{P}{V}$$

$$= \frac{1 \times 10^5 \text{ W}}{5 \times 10^4 \text{ V}}$$

$$= 2 \frac{\text{W}}{\text{V}}$$

$$= 2 \text{ A}$$

(ii) (a) $P = I^2R$

$$= (200 \text{ A})^2 \times 0.25 \ \Omega$$

$$= 1 \times 10^4 \text{ A}^2\Omega$$

$$= 1 \times 10^4 \text{ W}$$

(b) $P = I^2R$

$$= (2 \text{ A})^2 \times 0.25 \ \Omega$$

$$= 1 \text{ A}^2\Omega$$

$$= 1 \text{ W}$$

(iii) (a) % power loss $= \dfrac{1 \times 10^4 \text{ W}}{1 \times 10^5 \text{ W}} \times 100\%$

$$= 10\%$$

(b) % power loss $= \dfrac{1 \text{ W}}{1 \times 10^5 \text{ W}} \times 100\%$

$$= 0.001\%$$

Consider the unit analysis for part (i). Within the SI, the power is in watts when the current intensity is in amperes and the potential difference is in volts. (Since watts = amps × volts, then, at the division stage, we have $\dfrac{V \times A}{V}$, which gives amperes as the final unit.)

In part (ii) of each calculation, the SI unit of power (watts) results from the calculation involving the SI units of current intensity (amperes) and resistance (ohms). In the multiplication step, it may not be clear how $A^2\Omega$ is equivalent to W. From the relationship $V = IR$, you can see that $1 \text{ V} = 1 \text{ A}\Omega$. Therefore, $1 \text{ A}^2\Omega = 1 \text{ AV}$. But $1 \text{ A} = 1 \text{ C/s}$, and $1 \text{ V} = 1 \text{ J/C}$. So, $1 \text{ AV} = 1 \text{ C/s} \times 1 \text{ J/C} = 1 \text{ J/s}$, or 1 W.

You can see from the answers to parts (ii) and (iii) of the example how dramatic is the effect of varying the potential difference.

Increasing the potential difference by a factor of 100 decreases the power loss by a factor of 10 000! So now you know why electricity is carried from one place to another by high tension wires.

Questions

1. A cable of resistance 3 Ω carries 750 kW of power. If the line is at 75 kV, calculate:
   (a) the current intensity in the line
   (b) the power loss
   (c) the percent power loss

2. In a 5-Ω cable that carries 50 kW of power, 1% of the power is lost. By what factor must the potential difference be increased to reduce the percent power loss to 0.01%?

8.10 Energy and Power Problems

Now that you are aware of the relationships you used in the previous section, you can apply your knowledge to some more challenging problems.

Questions

1. In Example 1 of section 8.9, you saw that increasing the potential difference by a factor of 100 reduced the power loss by a factor of 10 000. In question 2 of that section, how did the potential difference need to change to give a 100-fold power loss decrease?
   (a) Can you generalize on the basis of these answers? In what way does the power loss depend mathematically on the potential difference in a certain cable?
   (b) If the current in a cable is doubled, by what factor will the power loss increase?

2. A 50-Ω electric lamp is connected to a 120-V supply for 40 min.
   (a) How much energy is used in
   (i) joules?
   (ii) kilowatt hours?
   (b) What is the cost of the energy if 1 kWh costs 4.03 cents?

3. Of the energy used in an incandescent light bulb, only 5% is converted to light. A 100-W bulb costs $1.32, and electric energy costs 4.03 cents per kWh.
   (a) If the bulb has a life of 800 h, what is the total cost of operation (including the cost of the bulb)?

(b) If 95% of the energy used by the bulb is converted to heat, how much heat does it produce in its life? Answer in:
(i) joules
(ii) megajoules
(c) What fraction of the total cost is the cost of the light bulb?

4. The power delivered by a high-tension wire whose resistance is 5 Ω is 200 kW. The power is transmitted at 75 kV. Determine:
(a) the current in the wire
(b) the power loss
(c) the percent power loss

5. The transmission lines used to distribute 750 kW of power have a resistance of 0.5 Ω. For a potential difference of:
(i) 750 kV (ii) 75 kV
calculate:
(a) the current intensity in the lines
(b) the power loss in the lines
(c) the percent power loss

6. In question 5(a) and (b), verify the units of your answer by means of unit analysis.

7. State the relationship between the increase in potential difference and the decrease in power loss in a wire.

8.11 Electrical Energy Used by a Resistor

In section 8.3, you learned how to calculate the energy consumption of an appliance whose power rating was known. But what if you do not know the power rating? How can you determine how much energy the appliance will use and the cost of running it? To answer these questions, you will now devise your own procedure.

Fig. 11 *This new type of light bulb saves energy.*

Materials

power supply
resistor (light bulb)
ammeter
voltmeter
connecting wires
stopwatch (or watch with a second hand)

Method

1. Devise your own procedure for measuring the quantity of energy used by the resistor in a certain period of time.

2. Draw a circuit diagram to represent your set up.

3. Have your suggestion checked by your teacher.

4. Carry out the procedure.

5. Write a report on your findings.

Follow-up

1. Calculate the quantity of electrical energy, in joules, used by the resistor in
   (a) 5 min   (b) 12 h   (c) 1 a

The unit a is short for "annum" (year).

2. Express your answers to follow-up question 1 in kilowatt hours.

3. Into what form(s) of energy did the resistor convert the electricity it used?

Investigation	*8.12 Transformation of Electrical Energy into Heat Energy*

You learned in a previous science course that, when solids absorb heat, they expand. Gaps between the ends of the steel rails on railroad tracks allow for expansion in summer. Bridges must be designed so that they do not buckle on hot days. Telephone wires sag more in the summer than in the winter, because they expand as the temperature increases. But how is the expansion of solids relevant to the study of electricity? The following demonstration will tell you the answer.

Materials

iron wire
2 stands
switch
power supply

Method

1. Stretch the iron wire between the two stands.

2. Connect the ends of the iron wire to the switch and power supply. Close the switch and observe the effect of the electric current on the wire.

3. Increase the current intensity and observe the effect on the wire.

Follow-up

1. Why did the length of the wire change?

2. Into what form of energy was the electricity transformed?

3. If you were to put up a new telephone cable in the summer, why would it not be a good idea to string it very tightly between two poles?

| Investigation | ## 8.13 Factors That Affect the Quantity of Heat Energy |

If you hold a spoon filled with water over a flame, the water may be raised to the boiling point in, say, 10 s. However, if you substitute a beaker that contains 200 mL of water, it may take 10 min to boil. The rise in temperature in the two cases is the same, but the quantities of heat used are very different. A few drops of hot water splashing onto you may do no harm, but a pailful of the same water could cause serious injury. Why? Discuss your ideas, then test them as follows.

Materials

hotplate
beaker
50-mL graduated cylinder
balance

2 large test tubes
thermometer
mineral oil
water

Method

I.

1. Pour 50 g (which is 50 mL) of water into one large test tube and 10 g (10 mL) of water into another. Measure the initial temperature of the water in both tubes.

2. Place both tubes in a beaker that is two-thirds full of water. Heat the beaker on a hotplate and measure the time it takes for each sample to reach 70°C. Record your findings.

II.

1. Measure 10 g (10 mL) samples of water into the two test tubes. Record the initial temperature of the water.

2. Place both tubes into a beaker that is two-thirds full of water and heat it on a hotplate. Record the time required for one sample to reach 50°C and the other to reach 70°C.

III.

1. Measure 10 g (10 mL) of water into one test tube and 10 g of mineral oil into the other. Record the initial temperatures.

2. Place both tubes into a beaker that is two-thirds full of water. Heat the beaker on a hotplate and record the time it takes for each sample to reach 70°C.

Follow-up

1. What was the manipulated variable in each part of the experiment?

2. State any relationships you can find that relate these variables to the quantity of heat absorbed by the liquids in the test tubes.

3. A common equation is used to calculate the quantity of heat, Q, absorbed by a sample of a substance that undergoes a temperature change. This equation is written as:

$$Q = mc\Delta T$$

where m is the mass of the sample, c is a characteristic property of the material known as its **specific heat capacity**, and ΔT is the temperature change.
   (a) Are the relationships you stated in question 2 consistent with this equation? Explain.
   (b) Use the equation to predict the units of specific heat capacity.
   (c) Suggest a definition for specific heat capacity.

4. Look up in a reference book the specific heat capacities of various substances. Determine whether the units and definition you proposed in question 3 were reasonable.

5. Look up the specific heat capacity of water. Use it to calculate the quantity of heat energy absorbed by 100 g of water to increase its temperature by 40°C.

6. Refer to the feature on James Watt and do some additional reading. Briefly summarize his role in the industrial revolution.

7. Write a report on this investigation. Include your answers to the follow-up questions.

ELECTRICAL TIDBITS

JAMES WATT

Watt made instruments for the University of Glasgow. His first important contribution was to design a steam engine. It had an efficiency of only about 2%. In other words, only about one-fiftieth of the energy of burning fuel could be converted into mechanical energy.

Watt made many improvements. Perhaps the most important was his invention of the double-acting engine. Steam was admitted by means of tubes and valves, first to one side and then to the other side of the piston. The steam pushed the piston backward as well as forward, so that the engine no longer relied on atmospheric pressure. His discoveries showed that heat was another form of energy, and could therefore be made to do work. The unit of power is named in his honour.

Fig. 12 *James Watt, 1736–1819*

Investigation	## 8.14 Conservation During Energy Transformations

You are aware that some electrical devices, such as electric kettles and irons, convert electrical energy into heat. But how does the quantity of electrical energy used by such a device compare with the quantity of heat it produces? Formulate a hypothesis and discuss it with your group. You will now answer the question by using a piece of apparatus called an electric calorimeter (Figure 13), which operates similarly to an electric kettle.

Fig. 13

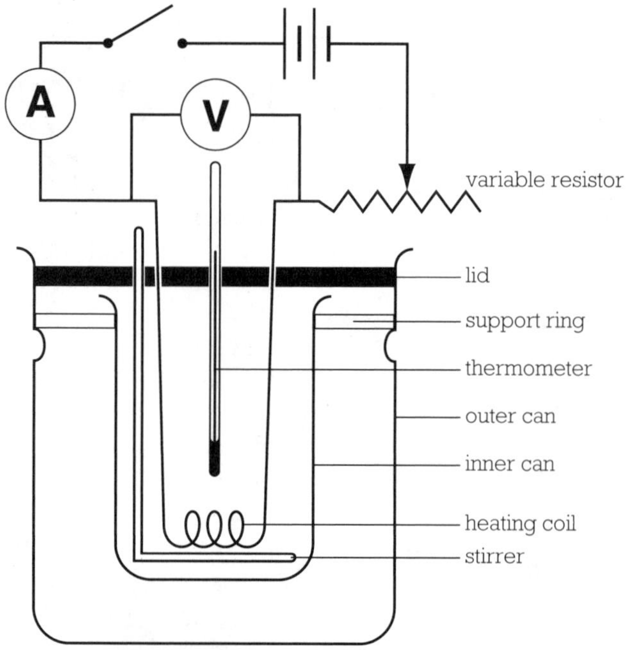

variable resistor
lid
support ring
thermometer
outer can
inner can
heating coil
stirrer

The calorimeter consists of two vessels, one inside the other. The inner vessel is a metal can of known mass and specific heat capacity. It is shiny to reduce the radiation of heat from within and the absorption of heat from without. The can is supported by a ring, made from a poor conductor of heat. The outer metallic container is also shiny. The air between the two vessels reduces losses or gains of heat by conduction.

The calorimeter has a wooden lid to reduce heat losses. This lid has two holes in it, one for a stirrer and the other for a thermometer. The lid is fitted with two terminals to allow the heating coil (a resistor) inside the inner vessel to be connected to a power supply. Thus the calorimeter provides a means of heating substances electrically under carefully controlled conditions.

Materials

power supply
electric calorimeter
ammeter
voltmeter
stopwatch (or watch with a second hand)
thermometer
balance

Method

1. Develop your own procedure for measuring the quantity of heat energy produced from a known quantity of electrical energy in a circuit. (Make sure that your procedure allows you to determine the quantities of both the electrical energy used and the heat energy produced.)

2. List all the physical quantities you need to measure.

3. Design a circuit diagram. Then discuss it with your group and have your teacher check it.

4. Conduct the experiment.

Follow-up

1. Calculate the electrical energy used by the calorimeter.

2. Calculate the heat energy produced in the calorimeter.

3. How do your answers to questions 1 and 2 compare? Can you formulate a law on the basis of the experimental results?

4. Discuss the findings of the entire class. Do other students agree with the law you formulated?

5. Describe the likely sources of error in your own result.

6. Do you think that energy is conserved in other processes? (Think of specific examples, such as the processes that occur inside your own body. What form(s) of energy are involved? Does "energy input" equal "energy output"?)

7. Research and describe some of the major discoveries of the physicist James Prescott Joule.

8. List at least four devices whose purpose is to convert electrical energy into heat energy.

9. Write a report on this investigation. Include the answers to all the follow-up questions.

8.15 Conservation Calculations

The idea that energy input equals energy output in a process (in other words, the idea that energy is conserved) is generally accepted. However, in practice, the energy output of a device is not always in the most desirable form. An incandescent light bulb, for example, may convert only 5% of the electrical energy it uses into light energy. The rest is converted into heat. Therefore, in terms of the useful energy it produces, we might describe the light bulb as only 5% efficient. (Of course, if we dreamed up a use for the light bulb as a heater, then we might think of it as 95% efficient!)

We can define percent efficiency as follows:

$$\text{Percent efficiency} = \frac{\text{useful energy output}}{\text{energy input}} \times 100\%$$

Example 1

An electric heater is 90% efficient and uses 20 kJ of electrical energy in a certain time period. How much heat energy does it produce in that period?

Solution

$$90\% = \frac{\text{heat produced}}{\text{electrical energy used}} \times 100\%$$

$$= \frac{Q}{20\,\text{kJ}} \times 100\%$$

$$\text{So} \quad Q = \frac{20\,\text{kJ} \times 90\%}{100\%}$$

$$= 18\,\text{kJ}$$

The heater produces 18 kJ of heat energy.

Example 2

A 1000-W electric water heater changes the temperature of 30 kg of water by 13°C in 30 min. If the specific heat capacity of water is 4.18 J/g°C, determine:
(a) the quantity of electrical energy used, in joules and kilowatt hours
(b) the cost of the electrical energy if 1 kWh costs 4.03 cents
(c) the quantity of heat energy absorbed by the water, in joules
(d) the percent efficiency of the energy conversion

Solution

(a) Electrical energy used, $E = Pt$

$$= 1000 \text{ W} \times 1800 \text{ s}$$
$$= 1.8 \times 10^6 \text{ Ws}$$
$$= 1.8 \times 10^6 \text{ J}$$

$1 \text{ kWh} = 3.6 \times 10^6 \text{ J}$

So $E = 1.8 \times 10^6 \text{ J} \times \dfrac{1 \text{ kWh}}{3.6 \times 10^6 \text{ J}}$

$$= 0.50 \text{ kWh}$$

(b) Cost $= 0.50 \text{ kWh} \times \dfrac{4.03 \text{ cents}}{1 \text{ kWh}}$

$$= 2.0 \text{ cents}$$

(c) Quantity of heat, $Q = mc\Delta T$

$$= 30\,000 \text{ g} \times 4.18 \text{ J/g}°\text{C} \times 13°\text{C}$$
$$= 1.6 \times 10^6 \text{ J}$$

(d) Percent efficiency $= \dfrac{\text{useful energy output}}{\text{energy input}} \times 100\%$

$$= \dfrac{1.6 \times 10^6 \text{ J}}{1.8 \times 10^6 \text{ J}} \times 100\%$$

$$= 89\%$$

Questions

1. The current through a 20-Ω heater is 6.0 A. Calculate:
   (a) its power rating
   (b) the heat energy it produces in joules if it is operated for 10 min. (Assume 100% efficiency.)

2. An experiment with an electric calorimeter gave the following data:
   resistance of coil $= 55.0 \ \Omega$
   potential difference across coil $= 110$ V
   mass of water in calorimeter $= 156$ g
   mass of calorimeter $= 60$ g
   specific heat capacity of calorimeter $= 0.418$ J/g°C
   specific heat capacity of water $= 4.18$ J/g°C
   time of run $= 2$ min
   temperature increase of calorimeter and contents $= 25°\text{C}$

Determine:
(a) the electrical energy consumed
(b) the heat energy produced
(c) the percent efficiency of the energy conversion

3. An electric heater used for heating small quantities of water operates at 110 V. It can heat 1.0 L of water from 20°C to 80°C in 10 min. Its efficiency is 80%. What is the cost of heating the water, if electrical energy costs 4.03 cents per kilowatt hour?

4. An experiment yields the following data:

mass of calorimeter = 20 g
mass of calorimeter plus water = 195 g
initial temperature = 19.0°C
final temperature = 23.0°C
elapsed time = 150 s
current = 5.0 A
potential difference = 4.0 V

If the specific heat capacity of the calorimeter is 0.836 J/g°C, calculate:
(a) the quantity of electrical energy used
(b) the quantity of heat energy produced
(c) the percent efficiency of the energy conversion

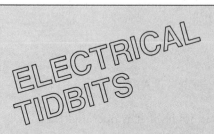

THE ELECTRIC ARC

By now, you are aware that different electrical devices have very different power requirements. An electric iron is much more powerful than a doorbell, for example. But, generally speaking, we think of electrical devices around the home as things we just plug in and turn on. We are most familiar with items that work at the potential difference of the household mains, 110-120 V, and draw less than 15 A of current.

We are aware of some devices that require lower potential differences. Obvious examples are flashlights and portable radios that operate on low-voltage batteries.

There are other types of electrical equipment that are outside our normal range of experience. One of these is an electric arc. This is produced when a potential difference of at least 45 V is applied to two carbon rods. The ends of the rods are then brought into con-

8.16 Energy Transformations in Consumer Goods

In previous sections, you have encountered examples of consumer goods that transform energy. The emphasis has been on the transformation of electrical energy into heat. However, the transformations can involve other forms of energy. Use the following questions to focus your attention on a wider variety of energy transformations in consumer goods.

Questions

1. Make a list of at least ten consumer goods that transform energy.

2. For each item you listed in question 1, indicate the form(s) of energy involved.

3. Indicate the energy transformation that each item is designed to bring about.

4. In each case, indicate any reasons for energy loss during the transformation.

5. In each case, suggest ways of reducing the energy lost during the transformation.

tact and slowly drawn apart. A brilliant arc of white light appears between the ends of the carbon rods. So much heat is generated that the rods become white-hot and begin to vaporize. The temperature reaches around 4000°C.

The carbon vapour in the gap between the electrodes emits light and helps conduct electricity across the gap. When the arc is in steady operation, the gradual vaporization of its carbon atoms tends to increase the size of the gap between them. However, an electric motor or clockwork mechanism is employed to keep the carbon rods the correct distance apart as they burn away.

Some electric arcs, such as those used in powerful searchlights and in some types of welding, require huge currents. But there are lower powered applications, such as laboratory arc lamps. Carbon arcs can operate on 110 V or 220 V of direct current. The ones used in motion picture theatres require from fifty to several hundred amperes.

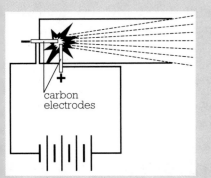

carbon electrodes

Fig. 14 *A simple arc lamp.*

| Investigation | ## 8.17 Electrical Energy Trans-formations |

You have considered the transformation of electrical energy into heat energy in considerable detail. You will now design an experiment to study the transformation of electrical energy into a form of energy other than heat. Before you begin, list some uses of electrical energy you are familiar with. Restrict your list to applications in which the desired form of energy output is not heat. Discuss your list with others in your group and choose a type of energy transformation you would like to study.

Materials

Develop a list in consultation with your teacher. Check what devices are available for transforming electrical energy into a form other than heat.

Method

1. Design a circuit diagram to represent the experiment you intend to carry out. Have your teacher check the diagram.

2. Assemble the circuit and carry out the experiment.

3. Examine each part of your circuit and record the energy transformations that are occurring.

4. Write a report that includes the answers to the follow-up questions.

Follow-up

1. You are unlikely to observe that all of the electrical energy used in the circuit is transformed into the form of energy you want.
   (a) Where in your circuit does most of the energy loss seem to occur?
   (b) In what form does most of the energy loss occur?

2. Suggest ways of minimizing the energy loss during your energy transformation.

8.18 Production of Electrical Energy

About 30% of the industrialized world's energy resources go to generate electricity. This proportion is likely to increase to 40% in this decade. The ways in which electricity is produced in different regions vary.

Electricity is a manufactured product. One of its limitations is that it cannot be stockpiled. It must be manufactured at the time it is needed. For this reason, electrical generating companies cannot produce electricity at a rate that satisfies the average demand. They must have sufficient capacity to satisfy peak demand. At times when local demand is lower, some of their facilities will be under-used, unless alternative markets are available.

Questions

1. Consult reference materials to find out how electricity is generated in your province.

2. Describe the energy transformation(s) associated with each method of producing electricity.

3. List the advantages and disadvantages of each of these methods of producing electrical energy.

4. Design a circuit you could use to produce electrical energy. Draw a circuit diagram.

8.19 Environmental Impact of Electrical Energy Production

It is no easy task to decide which methods of producing electricity are the most environmentally friendly. The first stage in making such a decision must be to gather information about energy technologies. You can then use this information to reach your own conclusions. A good deal of research will be necessary to do justice to the following questions.

Questions

1. Refer to the methods of generating electricity you listed in the previous section. Identify ways in which each method is beneficial or harmful to the environment.

2. Assess the overall environmental and social impact of each method. (This is complex and involves the kind of environmental issues you raised in question 1, as well as such social concerns as employment, working conditions, and so on.)

3. Suggest ways of minimizing the negative environmental impact of the production of electricity.

4. Assess the economic feasibility of your proposals in question 3.

8.20 Research into Energy Transformation Techniques

The production of electrical energy from falling water requires a different technology than does its production from burning fossil fuels, such as coal, oil, and natural gas. Each technology had to be developed sufficiently to permit a particular method of electricity production. Research continues into ways of improving existing technologies and developing new ones. Completion of the following questions will help you become aware of recent developments.

Questions

1. Use reference materials to become familiar with the state of research into the use of new sources of energy.

2. Predict how each of these new sources would affect the environment.

3. List the forms of energy that cannot be tranformed effectively into electricity with present technology.

4. Use reference materials to identify your province's contributions to the development of energy transformation techniques.

Investigation

8.21 Perpetual Transformation of Energy

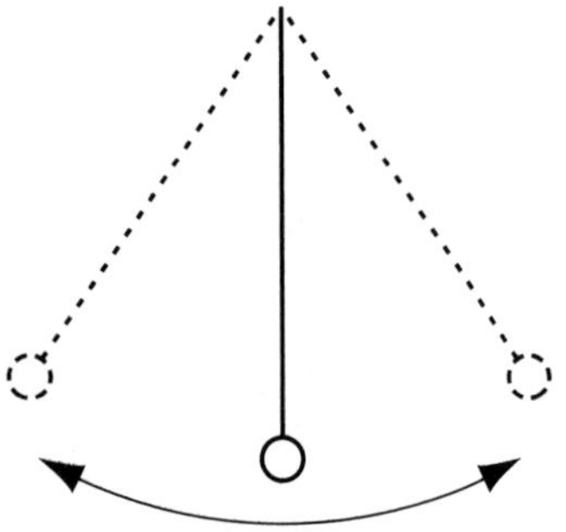

Consider a simple pendulum (Figure 15). A small ball is suspended from a fixed point by means of a string.

When the ball is pulled aside, it is raised a certain distance and hence given some potential energy. When the ball is let go, it moves toward its lowest point, and its energy changes from potential to kinetic. Energy is conserved in the process. However, some of the energy is converted into heat, because of friction at the point of suspension and friction with the air (air resistance). But what if there were no air resistance and no friction at the point of suspension? Then no heat would be produced, and the sum of the potential energy and kinetic energy would be constant at any point along the ball's path. The ball would keep swinging forever and

Fig. 15

would always rise to the same height at the end of each swing. This type of energy transformation is called "perpetual," meaning that it would go on forever. You have probably heard of the term "perpetual motion," a type of energy transformation in which the energy produced is kinetic. If the pendulum swung forever, that would be an example of perpetual motion.

The above example of a perpetual energy transformation would depend on a complete lack of friction during the swinging of the pendulum. Is this possible in practice? What is the possibility of bringing about other perpetual energy transformations? You will now examine whether you can devise a perpetual energy transformation that involves electrical energy.

Materials

Devise your own materials list in consultation with your teacher.

Method

1. Suggest a way of bringing about successive and perpetual energy transformations, based on the law of conservation of energy. (Make sure that electrical energy is involved. For example, you might consider an electrical-thermal-electrical sequence.)

2. Devise a procedure and draw a circuit diagram to represent your suggested method.

3. Have your teacher check your procedure, then carry it out.

4. Use reference materials to research the history of attempts to design a system capable of bringing about perpetual energy transformations.

5. What is the general opinion of scientists about the possibility of bringing about perpetual energy transformations?

6. You are familiar with many energy transformations that are not perpetual. Why are they not perpetual? (In other words, how is energy dissipated during these transformations?) Consider at least three specific examples.

8.22 Energy in our Daily Lives

As stated earlier, we take electrical energy for granted in many of the things we do. But how has our dependence on electricity come about, and how does our use of electricity compare with that in other parts of the world?

In the Third World, per capita use of commercially supplied energy is small compared with that of the industrialized countries. Even so, the size and rapid growth of the Third World population, along with the relatively inefficient use of energy in many developing countries, contribute significantly to the rapid growth in global energy use. Many people in the Third World rely on fuelwood for energy, yet many developing countries face shortages of this resource.

Patterns in world energy use are marked by striking contrasts and inequalities. An average Third World resident uses less than one-twelfth the energy consumed by an average Canadian resident, for example.

Questions

1. What are the definitions of the terms "physical change" and "chemical change"?

2. Draw up a list of changes from your everyday life that require energy.

3. Classify the changes you listed in question 2 as physical or chemical changes.

4. Research and distinguish the sources of energy used in an industrial society and those used in a non-industrial society.

5. Research and distinguish the quantities of energy used in an industrial and a non-industrial society.

6. Find out the total annual energy consumption of your province. From this information and the total population of the province, calculate the per capita energy consumption.

8.23 The Provincial Electrical Power Network

The supply of electrical energy in Canada is undertaken by provincial and municipal utility companies. Examples of large provincial organizations are Hydro-Québec and Ontario Hydro. To examine one of these in greater detail, Hydro-Québec was created by the Quebec government in 1944 through the acquisition of Montréal Light, Heat and Power Consolidated, and Beauharnois Light.

In 1963, Hydro-Québec acquired most of the remaining private electrical co-operatives in Québec and gradually established a vast, integrated electrical generating, transmission, and distribution system across the province.

Fig. 16 *Hydro-Québec gets some of its electrical power from Churchill Falls, Labrador.*

The following questions will prompt you to look at and think about the history and scope of your provincial electrical system in more detail. You will probably need to contact your provincial utility corporation to obtain information.

Questions

1. Identify the parts of the provincial power network you can see in your community.

2. Compare the size of the power network in your region with the overall provincial network.

3. Research the history of the development of the electrical power network in your region.

4. Compare the development of the power network in your region with the development of the provincial network.

5. Assess the environmental impact of the power network in your region.

6. Identify scientists, industrialists, and politicians who contributed to the development of the electrical power network in your province.

8.24 Science, Technology, and the Power Network

Science and technology are often interdependent. Many technological innovations develop from scientific knowledge, and vice versa. In their zest to discover the natural world around them, the ancient Greeks used their scientific knowledge to develop a new piece of technology — the telescope. As a result, new scientific fields could develop, for example, optics and astronomy. Scientific knowledge of the behaviour of light allowed the development of the microscope, which, in turn, opened up the field of microbiology. The discovery of electromagnetism led to the creation of new technology that changed the direction of the industrial revolution.

There are exceptions to the interdependence of science and technology. For example, it is unlikely that the people who invented one of the most important technological innovations — the wheel — understood the scientific principles that underlie it. Nevertheless, the interdependence of science and technology is strong.

The development of a provincial power network depends on the interaction of science and technology. Obviously, its development depends heavily on technological innovations, many of which depend on scientific discoveries. With a power network in place, new discoveries and inventions are possible.

Smelting is the process of melting an ore to get a metal out of it.

ELECTROTECHNOLOGY

An **electrotechnology** is a system or piece of equipment that uses electricity to produce or process consumer goods, or to carry out various industrial processes, such as heating, drying, and smelting. In the past, the energy needs for these tasks were met by fossil fuels, with electricity primarily being used for motors, lighting, and electrolysis.

Electrotechnologies now use electricity in most industrial processes. Many people regard electricity as more adaptable and more efficient than fossil fuels. In many cases, electrotechnologies lead to increased quality and productivity. They readily permit computerization and robotization of industrial processes (Figure 17).

Electricity is not necessarily made from fossil fuels, so it may be better for the environment. As Canada is a major producer and user of electricity, it is less vulnerable than some countries to uncertainties over the supply of fossil fuels.

What do you think of the development of more electrotechnologies and the increased consumption of electricity?

Fig. 17

Questions

1. What scientific discoveries contributed to the development of your province's power network?

2. What scientific discoveries and technological applications have resulted from the development of your province's power network?

3. In what ways did the discoveries and applications you listed in question 2 depend on the development of the power network?

8.25 Electricity and Daily Life

The consumption of electricity in the industrial world has increased enormously. To make this increased use possible, electrical networks have undergone great expansion. The following questions will help you develop your own knowledge and opinions on the increased use of electricity.

Questions

1. List domestic, commercial, and industrial uses of electricity (at least five of each).

2. Use reference materials to assess the increase in the production of electricity in your province over the past 50 years.

3. Relate the increase in the production of electricity with the uses identified in question 1.

4. How has the increase in the production of electricity affected life in your province?

5. How appropriate is your province's use of electrical energy?

6. Propose ways of making more effective use of your province's electrical energy.

8.26 Hydro-Electric Research

Some Canadian utility corporations are leaders in the production of hydro-electricity, that is, electricity produced from moving water. The following questions will give you the opportunity to examine the research activity in this field in your province and will allow you to examine its international impact.

Questions

1. Use reference materials to identify the latest hydro-electric research conducted in your province.

2. Identify the high-technology hydro-electric equipment from your province that is used abroad.

3. Identify firms and institutions in your province that have contributed to the development of hydro-electric power facilities in other parts of the world.

4. Describe foreign hydro-electric power facilities built by people from your province.

5. Give examples of the role of your province in training representatives from other countries to become experts in hydro-electric production.

P O I N T S · T O · R E C A L L

- Energy is the ability to do work. It is measured in joules (J).
- Electrical energy is also measured in kilowatt hours (kWh).
- Power is the rate of using energy. It is measured in watts (W):
$$\text{Power} = \frac{\text{energy}}{\text{time}}$$
- A watt is the power when one joule of work is done in one second.
- Electrical energy is the energy carried by moving charges. (In a conducting wire, the charges are carried by moving electrons.)
- The potential difference across two points in a circuit is measured in volts (V).
- Electric charge is measured in coulombs, symbol C.
- When one joule of work is done to separate one coulomb of charge, the potential difference is one volt.

- Power (watts) = potential difference (volts) × electric current (amperes)
- Electrical energy is transmitted at high potential difference and low current intensity to minimize heat loss during transmission.
- The loss of electrical energy in the form of heat is known as the Joule effect.
- Energy is conserved during energy transformations.
- Percent efficiency = $\dfrac{\text{useful energy output}}{\text{energy input}} \times 100\%$
- Perpetual transformation of energy is the transformation from one form to another without any loss of energy.
- The quantity of heat absorbed by a substance is given by $Q = mc\Delta T$.

R E V I E W · Q U E S T I O N S

1. For each appliance listed, calculate the quantity of energy used, in kWh, and the total cost. Assume that the cost per kilowatt hour is 4.03 cents.
   (a) 1200-W coffee percolator used for 3.0 h
   (b) 1.5-kW oven used for 40 min
   (c) 60-W light bulb turned on for 6.0 h

2. The diagrams in Figure 18 show electric meter readings taken one month apart. Calculate the cost of electrical energy for the month if the unit cost is 4.03 cents per kilowatt hour.

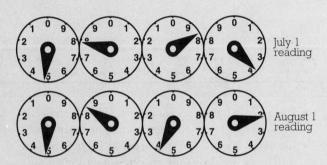

July 1 reading

August 1 reading

Fig. 18

3. How much electrical power is used to light a room with a six-lamp chandelier, if each lamp uses 0.4 A from a 110-V power line?

4. How much energy, in joules, is used when a 100-W lamp is turned on for 30 min?

5. A dryer is rated at 5000 W, and a colour TV at 300 W. The colour TV uses more electrical energy in a month than the dryer. Explain.

6. A transmission line has a resistance of 5 Ω. Calculate the power loss in the line when 10 000 kW are transmitted at:
   (a) 10 kV   (b) 100 kV

7. What is meant by the term "specific heat capacity."

8. An experiment to measure the electrical energy tranformed into heat by an electric calorimeter gave the following data:

   resistance of the coil = 55 Ω
   potential difference across the coil = 110 V
   mass of water = 156 g
   mass of calorimeter cup = 60 g
   specific heat capacity of water = 4.18 J/g°C
   specific heat capacity of calorimeter cup = 0.418 J/g°C
   time of use = 1.25 min
   temperature increase = 23°C

   Determine:
   (a) the electrical energy used
   (b) the heat energy measured
   (c) the loss of energy, if any

9. List at least five ways in which your electrical energy use at home could be reduced.

10. Explain how to connect 1.5-V cells to obtain a potential difference of 6.0 V. Draw a diagram to show the arrangement.

11. Why is electrical power transmitted at high voltage?

12. A water bed heater has a resistance of 40 Ω. What is its power rating if it is operated on a 120-V circuit?

13. List several primary energy sources that are converted to electrical energy.

14. An electric iron has a mass of 1.0 kg. The heating element is made of nichrome wire. What time is required for the iron to be heated from 20.0°C to 130°C, if its power rating is 1100 W? Assume no loss of heat. (Iron has a specific heat capacity of 0.45 J/g°C.)

15. An electric hotplate draws 9.0 A on a 110-V circuit. In 6.00 min, the hotplate can heat 1.00 kg of water from 20.0°C to boiling. Calculate:
    (a) the energy loss
    (b) the percentage energy loss

16. If electrical energy costs 4.03 cents per kilowatt hour, what is the cost of heating 4.0 kg of water from 25°C to 95°C, assuming no loss of energy?

17. A student is planning to make an electric heater. The wire available has a resistance of 5.2 Ω/m. What length of this wire is needed to make a heating element that will draw 8.2 A from a 110-V power line?

18. A heating coil is connected to a 110-V circuit. The coil draws 8.0 A. How much energy, in kilojoules, is produced if the heater operates for 35 min, assuming no loss in energy?

19. Starting from oil, what steps would you take to convert chemical energy into electrical energy?

20. A hydro-electric plant may not cause thermal pollution, but it may cause other environmental problems. List at least three of them.

21. A high-tension transmission line passes through your community. Describe its possible health effects on humans.

22. Why is nuclear fusion not a commercially viable source of energy at the present time?

23. What is meant by the term "thermal pollution"? Describe its effects on air and water.

24. Identify three alternative energy sources for conversion to electrical energy.

9

INTRODUCING ACIDS, BASES, AND SALTS

Solutions of acids, bases, and salts are very important in chemistry.

CONTENTS

9.1 Research Project

In this chapter, you will learn a good deal about types of substances called acids, bases, and salts. When you have sufficient background information about these types of substances, you will carry out a research project that relates to at least one acid, base, or salt. The knowledge of research projects that you gained in sections 1.3 and 6.1 will help you. It is probably best not to read the rest of this section until your teacher tells you to.

From the following list, select a topic that interests you. All the topics concern acids, bases, or salts that you meet in your environment. Once you have selected a topic, tell your teacher. Your teacher can suggest when you have learned enough about acids, bases, and salts to tackle your chosen topic. When you are ready to begin, explore the topic, using the resources available to you — people, libraries, films, and so on, as you did in section 1.3.

Any solution of a salt is known as **saline**.

- determining the effect of antacids on one or more acidic solutions
- comparing the effectiveness of various antacids
- measuring the pH of precipitation
- monitoring the pH of snow over the course of the winter
- determining the effect of acidic, basic, neutral, and saline solutions on hair
- determining the effect of acid precipitation on various building materials
- determining the effect of acidic, basic, neutral, and saline solutions on various building materials
- comparing the acidity of various shampoos and toothpastes

253

- determining the turning point of a particular mixture of indicators
- determining the effect of acidic, basic, neutral, and saline solutions on plant growth
- determining the effect of ionic and covalent solutes on the melting point of a solution
- showing how an ionic solute and a non-ionic solute of equal concentration affect the freezing point of a solution
- determining how the pH of soil is affected by the addition of eggshells, lime, a hydrogen carbonate, or another substance
- comparing the pH values of commonly used substances
- measuring the quantity of an acid or a base in a household product in solution, in a food, or in another substance by means of the titration procedure
- determining the effect of a base on animal and vegetable fats, and mineral oils

Perhaps you are interested in doing a different project about acids, bases, and salts. If so, discuss your ideas with your teacher.

Make sure you have a clear idea of the goal of your research project. Be able to state your goal as a simple question.

Once you have decided on a topic, make lists of what you already know about this topic and your sources of information. What are some of the possible answers to your question? Which one makes the most sense to you? Select this answer to be your hypothesis. When the project is finished, you should know whether your hypothesis was correct.

As you already know, a successful research project depends on good organization. The following steps will help you.

Remember that to hypothesize means to suppose.

Planning

1. List all the tasks — even the smallest ones — that you must do in this project.

2. Know what measurements you will need to take and how you will take them. Are there any conditions, for example, temperature or concentration, that you may want to keep constant? What conditions will you vary so you can observe their effects?

3. Plan how to record your data. Draw up a data chart.

4. List all the equipment and chemicals that you will need. Draw a diagram of the apparatus assembly, if any, that you plan to use.

5. Have your teacher approve all of your plans before you begin your project.

Experimenting

1. Set up the apparatus.

2. Carefully take measurements and observe changes. Record all your measurements and observations as soon as you make them.

3. Compare your results with your hypothesis.

4. If the results are not what you expected, improve your hypothesis as you continue working on your project.

5. Test the improved hypothesis to see if it is better than your last one.

Observe all the safety precautions that you have learned, especially the wearing of safety goggles.

Presenting Your Results

A well-written research report must allow any reader to understand how the goal of the project was chosen, how you planned the project, and what you discovered.

1. *What you set out to do:* In the first part of your report, describe your topic. Include your hypothesis with this description. It does not matter whether your hypothesis turned out to be correct. What is important is that you learned something from testing this hypothesis. Never forget that scientists are always working at the limits of their knowledge, and there are many questions for which they do not have answers. If scientists could predict with 100% accuracy all of the answers, then there would be no new discoveries in science.

2. *How you carried out your experiment:* In the second part of your report, briefly tell the reader how you conducted your experiment. List any equipment and substances you used. If you assembled any apparatus to do your experiment, include a diagram of the assembly. (Keep in mind that your reader may want to assemble apparatus in the same way, so make sure your diagram is legible.)

   Record all your observations. Be the ears, eyes, and fingers of your readers. Write as if you were reporting an event over the radio and your audience could only imagine what you were experiencing.

   If you have any quantitative observations, record them in a chart or a graph. This will help your reader to recognize patterns and trends in your observations.

3. *What you discovered:* In this third part, tell your reader what you learned from your research and experiment. Keep in mind that you have a question to answer—what is this answer? Why do

you think that is the best answer? Does the answer agree with your hypothesis? If not, how can you modify your hypothesis? How does the answer to your question change the way you think about the universe or the way you live? Does it make you wonder about the answers to other questions? If so, what questions?

4. *Your sources:* List all the sources of information you used in your project.

9.2 Found in the Strangest Places

Our word "salt" shares a history with the word "salary." Both come from the Latin word *salarium*. Roman soldiers were paid with salt, which they used to preserve food.

The word "acid" comes from the Latin word *acidus*, meaning sour. Think of the sour taste of lemon and vinegar, both of which contain acids.

Acids, bases and salts are part of our daily lives. The orange juice we drink contains citric acid. The vinegar on our French fries is acetic acid. Soap is made using sodium hydroxide, a base. This base is also used to manufacture the paper in this book. We flavour our food with table salt, sodium chloride, and we melt the ice on our streets and sidewalks with another kind of salt, calcium chloride.

Figures 1 to 12 are pictures relating to acids, bases, and salts in everyday life. Can you match them with the captions a) to l) below? First, write the figure numbers in the margin of your notebook. Then beside each number, write the letter of the caption that fits best. The first one has been completed.

(a) *Fertilizers often contain phosphoric acid.*

(b) *It takes less work to beat egg whites until they are stiff if acid is added.*

(c) *The dark colour of chocolate cakes depends on how much base is added to the batter.*

(d) *Indigestion is caused by too much acid in the stomach. You can relieve the pain by taking an antacid, which is a base that destroys some of the acid.*

(e) *A car battery produces enough electrical energy to start the engine. One of the chemicals in a car battery is sulfuric acid.*

(f) *Acid rain causes millions of dollars' worth of damage each year to the outsides of buildings, to marble statues, and to the environment.*

(g) *Lactic acid in milk will stain the enamel surfaces of appliances.*

(h) *Pearls may be damaged by perspiration, which is acidic, and by acidic foods or drinks.*

(i) *Solid drain openers, which contain a base, clear blocked drains and sinks.*

(j) *Farmers add ammonia, a base, to soil so that crops will grow well.*

(k) *Silver nitrate is a light-sensitive salt that has been used in the photographic industry since the early days of glass plates.*

(l) *Plaster of Paris is a salt that forms a hard but brittle protection for broken bones.*

Fig. 1 (j)

Fig. 2

Fig. 3

Fig. 4

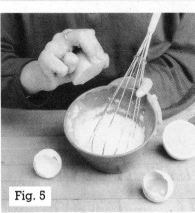

Fig. 5

Fig. 6

Fig. 7

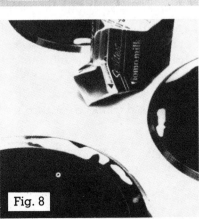

Fig. 8

Fig. 9

Fig. 10

Fig. 11

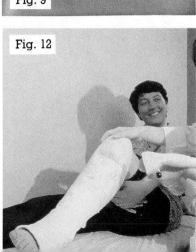

Fig. 12

9.3 Common Sense With Acids and Bases

"Alkali" is another word for base. In ancient times, the kali plant was burned to produce a basic ash. Bases such as soaps feel slippery and have a bitter taste. (Remember never to taste any chemicals in the laboratory.)

Fig. 13 *Never add a hypochlorite bleach, such as Javex, to a toilet bowl that already contains a toilet bowl cleaner. Toilet bowl cleaners are acidic and react with hypochlorite bleach to produce the poisonous gas chlorine. For safety's sake, do not mix bleaches with other chemicals.*

As you can see, acids and bases have many uses and are found in some unexpected places. Some acids, such as vinegar, and some bases, such as antacid tablets, are very safe to handle under normal circumstances. But others, such as the acid used in car batteries and the base used for clearing drains, must be stored and handled very carefully. Here are some examples of what can happen when acids and bases are used or stored incorrectly.

- A child was rushed to hospital after swallowing liquid drain cleaner. An investigation showed that the liquid had been stored in a soft drink bottle under the kitchen sink. The child almost died, and doctors had to reconstruct his damaged esophagus. His recovery took many months.

- A mechanic dropped a car battery. It cracked and the sulfuric acid splashed into the mechanic's eyes. The mechanic suffered eye damage.

- There is a story that the Emperor Nero played his fiddle while Rome burned. Some of the Roman nobility at that time showed signs of mental problems. Some scientists believe that the lead used in Roman times to make water pipes and wine goblets caused some of these mental problems. It is known that lead slowly reacts and dissolves in some water solutions, especially when the solutions are acidic. Today, we avoid storing acidic foods and drinks in lead containers, or in pottery with lead in the glaze.

- If you spray an oven-cleaning product into a hot oven without using rubber gloves or protecting your face, you may start to cough, the skin on your hands and face may begin to sting and itch, and your eyes may hurt badly. These symptoms would be caused by the presence of powdered base in the cleaner.

- A cook found that when canned pineapple was mixed with strawberries, the berries turned an unpleasant blue colour. However, if the cook used fresh pineapple instead, this change did not occur. The acid from the pineapple had dissolved iron from the cans, and the iron had turned the red colour in berries blue!

Questions

1. List all the foods — like lemon juice and vinegar — that taste sour. What kind of chemical substance do they likely contain?

2. (a) Give three examples of foods or drinks you would avoid storing in lead containers or in pottery with a lead glaze.
   (b) Explain why acidic foods or drinks should not be stored in these types of containers.

3. (a) Why should you never store dangerous chemicals in soft drink bottles?
   (b) Where and how should such chemicals be stored?

4. It is most important to read labels on product containers. Why?

5. How can you avoid being hurt when you handle a car battery or an oven cleaner?

6. List ten safety precautions and rules that all homes should have. Do not restrict your list to the safe use of acids and bases.

7. (a) Based on your observations of everyday acids and bases, suggest definitions for the terms "acid" and "base."
   (b) What basis, if any, do you have for defining the term "salt"? If you feel you can define this term, state your definition.

Investigation | 9.4 Learning About Indicators

To make the best and safest use of acids, bases, and salts, we must be able to identify them. There is a simple way of doing this using certain chemicals, called **indicators**. Indicators change colour

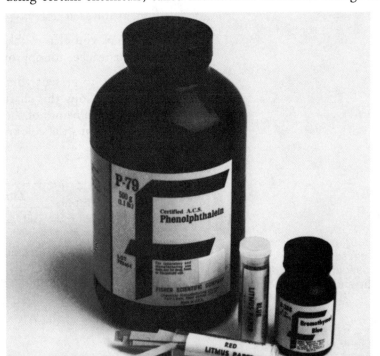

Fig. 14 *Some indicators: litmus, bromthymol blue (also known as bromothymol blue), and phenolphthalein.*

depending on the amount of acid or base in a solution. Let's now test some known acids, bases, and salts with certain indicators to discover the changes in colours.

One important acid–base indicator is called **litmus**. Papers containing the litmus dye come in two varieties—red and blue. To use litmus paper, just dip a piece into the solution to be tested. Another indicator, phenolphthalein, is dissolved in alcohol. To use this indicator, simply add a drop of it to the solution you are testing.

Materials

litmus paper (blue and red)
phenolphthalein solution in dropper bottles
known acids: vinegar (acetic acid); citric acid; muriatic acid (hydro-
    chloric acid)
known bases: filtered limewater (calcium hydroxide solution); dilute
    lye (sodium hydroxide solution)
known salts: table salt (sodium chloride); de-icing salt (calcium
    chloride)

As you develop a procedure, list in your notebook any equipment you need.

Method

1. Plan how you will test each solution with each indicator (Figure 15). Remember that you can only get accurate results if you add just *one* indicator to a solution at a time and you use clean equipment for each test.

2. List the steps you plan to take in this investigation. Show the list to your teacher for approval before you begin.

3. The best way to present your data is in a chart. Below is a sample chart. Copy this chart into your notebook and list in the first column the names of the solutions you are testing. Note in the other columns the colours you observe in the test with each indicator.

pH paper

solution

Fig. 15

Do not forget your safety goggles because the chemicals in this investigation can irritate your skin and damage your eyes.

Solution	Litmus		Phenolphthalein
	red	blue	

Follow-up

1. List all the solutions that turned blue litmus paper red and then all the solutions that turned red litmus blue.

2. (a) To what category — acid, base, or salt — do all the substances that turned blue litmus paper red belong?
   (b) To what category do all the substances that turned red litmus blue belong?

3. You may have noticed that some substances did not change the colour of litmus paper. These substances are said to be **neutral**. List all the neutral substances you tested.

4. (a) List all the substances that turned phenolphthalein pink.
   (b) How does this list compare to the lists you have already compiled?

5. Suggest an easy test to determine whether a solution contains an acid, a base, or a salt.

6. On the basis of your observations in this investigation, can you give new definitions for the terms ''acid,'' ''base,'' and ''salt''?

| Investigation | ## 9.5 *Getting Your Greens Really Green!* |

The correct chemical name for baking soda is sodium hydrogen carbonate. It is often called sodium bicarbonate.

Acids and bases change the appearance and texture of certain foods. Have you ever wondered why green vegetables sometimes look or feel awful when they are cooked? Let us investigate this by cooking some leafy green vegetables in baking soda, a common household base, and in vinegar, a common household acid.

Materials

safety goggles
three 250-mL beakers
teaspoon
tongs or fork
distilled water
5 mL sodium hydrogen carbonate (baking soda)
50 mL vinegar
spinach, cabbage, broccoli, or other leafy green vegetable

Method

1. Copy the following chart into your notebook.

	With water	**With base**	**With acid**
Colour			
Texture (feel)			

2. Prepare three 250-mL beakers as follows:
   (a) Add 100 mL of distilled water to beaker 1.
   (b) Add 1 teaspoon of sodium hydrogen carbonate to 100 mL of distilled water in beaker 2.
   (c) Add 50 mL of vinegar to 50 mL of distilled water in beaker 3.

3. Bring the liquid in each beaker to the boil, and add a leaf of spinach or of another green vegetable to each beaker. Cook for three minutes.

4. Remove the vegetables with tongs or a fork. Cool on a paper towel. Then observe the colour, and test the texture by rubbing each sample between two fingers.

Follow-up

1. Tap water in some homes may be basic because of a high mineral content in the water. How can a cook overcome this problem when cooking green vegetables?

2. Would you add lemon juice when cooking green vegetables? Explain your answer.

Chemically Speaking

HERE'S THE SECRET!

The results with the base probably reminded you of the joke: "First the good news and then the bad news!" Although the colour is greener than that of the vegetable cooked in vinegar or water, the base destroys the cell walls of the vegetable, making it too mushy to hold on a fork.

The cells of green vegetables contain green chlorophyll and yellow carotene. The colour of the chlorophyll dominates, so the vegetables appear green. The cells also contain acids. During cooking (in water or acid), the acids are set free and destroy some of the green chlorophyll but not the yellow carotene. The combination of yellow carotene and the remaining green chlorophyll gives an unappetizing olive green colour.

The best way to keep green vegetables green and appetizing when cooking them is to steam them. Place them in a container such as a sieve, colander, or steamer *above* boiling water. Do not cover them for the first part of the cooking. Any acids from the plant cells either escape in the steam, or drip into the water below.

| Investigation | 9.6 Chemical Properties of Acids and Bases |

In this investigation you will see if acids, bases, and salts will react with certain common materials. We will compare typical acids (hydrochloric and acetic acids) with typical bases (sodium hydroxide and potassium hydroxide) and typical salts (sodium chloride and sodium nitrate).

Materials

safety goggles
test tubes (at least 3)
test tube rack
wooden splints
long glass dropper
spatula

Use the acid and base with care. Wear your safety goggles.

Fig. 16 *In 1766, Henry Cavendish proved that the gas obtained when metals were added to acid solution was an element and produced only water when burned in air. The name hydrogen means "water maker."*

Recall from investigation 1.10 the lighted splint test used to identify hydrogen. Remember that hydrogen is flammable.

1 mol/L sodium hydroxide solution
1 mol/L potassium hydroxide solution
1 mol/L hydrochloric acid
white vinegar (acetic acid solution)
1 mol/L sodium chloride solution
1 mol/L sodium nitrate solution
magnesium ribbon
sodium hydrogen carbonate (baking soda)
calcium carbonate (marble chips)
filtered limewater
litmus paper (red and blue)

Method

Draw up a chart before you begin. Record all observations in the chart.

Chemicals used	Observations

I. *The Reaction With Litmus*

1. Test each of the solutions of sodium hydroxide, potassium hydroxide, hydrochloric acid, vinegar, sodium chloride, and sodium nitrate with litmus paper.

2. Classify these six solutions as acids, bases, or salts.

II. *The Reaction With a Metal*

1. Observe what happens when you add 2 or 3 mL of each solution to separate small samples of a metal, such as magnesium ribbon. If you see any bubbles forming, you will know that a gas is forming and escaping from the solution.

2. To test for hydrogen gas, bring a lighted splint to the mouth of the test tube.

(a) Which of the solutions reacted with the magnesium ribbon? Describe the reactions you observed.

(b) What happened in the lighted splint test and what did this test establish?

(c) The reaction of magnesium with hydrochloric acid is a single displacement reaction. Complete the word equation for the reaction.

magnesium + hydrochloric acid → _____ + _____

(d) What might happen if you spilled soft drink on the metal surface of your car and left it there?

(e) Why do soft drink manufacturers line aluminum cans with plastic coating?

(f) Does testing with magnesium ribbon allow you to distinguish among acids, bases, and salts? Explain.

III. *The Reaction With a Hydrogen Carbonate or a Carbonate*

1. Observe what happens when you add 2 or 3 mL of each solution to separate pea-sized quantities of solid sodium hydrogen carbonate.

2. Capture any gas that is formed by drawing it into a long glass dropper, which is placed just above the reaction mixture (Figure 17). Remove the dropper from the test tube.

3. Place the open end of the dropper below the surface of the limewater in a second test tube. Squeeze the bulb of the dropper to bubble the gas through the limewater (Figure 18). Carefully observe what happens to the limewater.

4. Repeat steps 1 to 3 using two marble chips (calcium carbonate) instead of sodium hydrogen carbonate.

The reaction of a carbonate or hydrogen carbonate with an acid produces carbon dioxide gas.

The word equations for two of the reactions you just performed are:

sodium hydrogen carbonate + hydrochloric acid →
         sodium chloride + water + carbon dioxide

calcium carbonate + hydrochloric acid →
         calcium chloride + water + carbon dioxide

(a) Which solutions reacted with the sodium hydrogen carbonate and the calcium carbonate? Describe the reaction that occurred.

(b) What happened in the limewater test and what did this test establish?

(c) Why is an acid mixed with sodium hydrogen carbonate in baking powder?

(d) Yogurt, sour cream, and chocolate are all somewhat acidic. Why is baking soda (sodium hydrogen carbonate) used to bake cakes that contain one of these ingredients? Why is it not necessary to use baking powder in this case?

(e) How does acid rain damage marble statues?

(f) Does testing with a hydrogen carbonate or a carbonate allow you to distinguish among acids, bases, and salts? Explain.

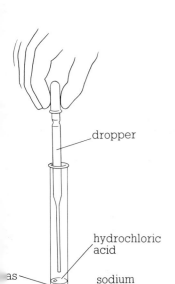

Fig. 17 *Capturing the gas produced.*

gas — dropper

hydrochloric acid

sodium hydrogen carbonate

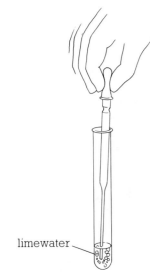

limewater

Fig. 18 *The limewater test is used to identify carbon dioxide gas.*

9.7 Acid Versus Base

Fig. 19 *Finger bowls of warm water and lemon slices are used to remove the odour of shellfish. Why do restaurants not provide finger bowls of vinegar?*

As you have learned, a neutral solution is neither acidic nor basic. In fact, an acidic solution can be used to destroy a basic solution. This process is called **neutralization** and is often used in our everyday life.

Some restaurants provide finger bowls of warm water and lemon juice (Figure 19). People may rinse their fingers in the water after eating shellfish. The chemicals that leave a fish-like odour on skin are bases. The acids in the lemon juice neutralize the bases and this removes the odour.

Bases damage animal fibres. However, years ago, many shampoos contained soaps, which are basic. These shampoos cleaned hair well but were harmful if left on the hair for too long. To neutralize any base left on the hair, people often used an acidic rinse, such as lemon juice or an acidic hair conditioner, after shampooing. In the same way, if a little vinegar or lemon juice is added to the rinse water when wool and silk are washed, any base that might otherwise damage the fabric will be neutralized.

SMOKING AND NEUTRALIZATION

A smoker craves a cigarette most after a meal, when drinking alcohol, or when under stress. Smokers obtain poisonous nicotine from the cigarette smoke they inhale. The inhaled nicotine is first absorbed into the lungs. It is then transported throughout the body in the bloodstream, reaching the brain in about 7 s. Some of the nicotine is stored in the brain.

The acidity of the blood increases after a person eats a meal, drinks alcohol, or is under stress. Nicotine is a base. The more acidic the blood, the more soluble the nicotine is in the blood. If the nicotine that is stored in the brain becomes dissolved in the blood, then the person starts to crave more nicotine.

Investigation | ## 9.8 A Simple Neutralization

If you neutralize a base with an acid, you may use an indicator to determine at what point the neutralization has taken place. When you see the final colour change, you will know you have added just enough acid to neutralize the base. Keep this in mind when you plan

Do you remember what colour phenolphthalein is in a basic solution? in an acidic solution? in a neutral solution? (Refer to investigation 9.4.)

Don't forget your safety goggles!

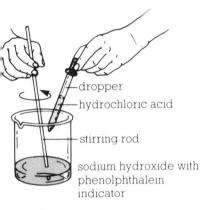

—dropper
—hydrochloric acid

—stirring rod

sodium hydroxide with phenolphthalein indicator

Fig. 20

how you will neutralize a sample of base with some acid. For example, you might want to add some phenolphthalein—one drop is enough—to 10 mL of base, and then you will be able to see when all the base has disappeared from the solution. Now find out exactly how much of the acid solution you need to destroy all the base.

Materials

Consider what you will need besides the acid, the base, and the phenolphthalein. First, you will measure the volume of base. Then add the acid gradually, drop by drop, until you can tell that the base has been completely neutralized.

Write down a list of the chemicals and the equipment you will need and have your teacher check it.

Method

1. Make a list of all the steps in your investigation.

2. Get your teacher's approval before you begin.

Follow-up

1. What happened when the acid was added to the base solution?

2. In question 1, how do you explain what happened? Test your explanation by using litmus paper.

3. List the types of materials you have investigated so far that react with each of the following.
   (a) acids
   (b) bases

4. If you spill some acid in the laboratory, what could you use to neutralize the acid?

Investigation

9.9 The Conductivity of Acidic, Basic, and Saline Solutions

You are aware that metals are excellent conductors of electricity. Electrical conductivity is an important physical property of all metals, but metals may not be the only substances that can conduct electricity.

You have just studied how acids, bases, and salts behave in the presence of indicators and with a variety of materials. Now you will see how these substances behave when an electric current passes through them.

In investigation 5.1, water was decomposed into hydrogen gas and oxygen gas. Do you recall that in step 1 you added some sulfuric acid to the water? Why do you think the sulfuric acid was necessary? Discuss your answer with your group.

Materials

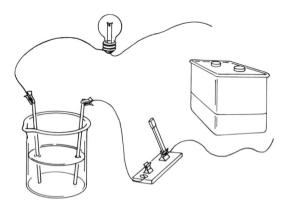

Fig. 21

Don't forget your safety goggles!

In order to test conductivity, you will need an apparatus similar to the one you used in investigation 1.10. You will dip the wire electrodes into various solutions. Be careful not to attach the wiring to the battery until after you have set up the test as shown in Figure 21.

You will need some solutions of acids, bases, and salts to test. You can test the acids you have met already — for example, acetic acid and hydrochloric acid — or you can test a solution of citric acid. The bases you have used already include potassium hydroxide and sodium hydroxide. Also, choose a salt or two.

Make a list of the chemicals and equipment you will need. Get your teacher's approval before continuing.

Method

1. List the steps necessary to test the conductivity of the solutions you have chosen. Remember that the metal electrodes must be washed with distilled water between tests! Show your method to your teacher for approval.

2. Prepare a chart to record the data. You may choose to use a chart similar to the one that follows.

Solution	Observation	Conductivity: yes or no?

Follow-up

1. How good was each solution at conducting electricity?

2. Were some solutions better conductors than others? What was the basis for your judgment?

Investigation

9.10 How Acids, Bases, and Salts Conduct Their Business

So far we have seen that acids and bases are quite powerful chemicals that react in a variety of interesting ways. Let's now investigate *why* these chemicals behave the way they do.

Recall that the solvent is the major component in a solution and that the solute is the minor component. Many of the solutions you have studied were prepared by dissolving a solid solute in water. In the conducting solutions of acids, bases, and salts, do you think that it was the solute, the solvent, or the mixture of the two that was responsible for the conductivity? Discuss your answer with the members of your group. Then find out whether your answer is correct by comparing the conductivity of pure water and of pure, dry acids, bases, and salts to the conductivity of the solutions.

Any solution in which the solvent is water is called an **aqueous** solution.

Materials

Since you will need a solid acid, you might consider testing citric acid, which is readily available in pure, solid form. For a solid base, you could use sodium hydroxide pellets, but be especially careful with them. If you need sodium hydroxide solution, your teacher will dissolve some of the pellets in water for you. Choosing a salt should be easy by now.

What sort of equipment will you require for testing conductivity? What glassware do you need? Since you cannot touch the

Don't forget your safety goggles!

Do not touch sodium hydroxide pellets with your hands because the pellets are very caustic.

sodium hydroxide pellets with your hands, how will you handle them? Make a "shopping list" of all the things you need and ask your teacher to approve it.

Method

1. Plan an experiment that will help you find out if it is water alone, the solute alone, or the mixture of the two that is responsible for conducting electricity.

2. Present your detailed plan to the teacher for approval.

3. Prepare a chart to record your data.

Follow-up

1. (a) Did any of the solutes conduct electricity by themselves?
   (b) Does distilled water conduct electricity?

2. Under what conditions do compounds such as citric acid, sodium hydroxide, and sodium chloride conduct electricity?

3. In general, under what conditions do compounds such as citric acid behave like acids?

4. Under what conditions do compounds such as sodium hydroxide behave like bases?

5. (a) Check the hypotheses you proposed in questions 3 and 4 by testing the dry acids and bases as well as their solutions with litmus paper.
   (b) Are your hypotheses correct?

6. (a) What is an electric current?
   (b) Suggest why acids, bases, and salts dissolved in water conduct electricity. Compare your explanation with those of other students.

9.11 Identifying Acids, Bases, and Salts From Their Formulas

The molecular formulas of some acids that you have used in your investigations are listed below. Do you notice anything that their formulas have in common?

Hydrochloric acid	HCl
Sulfuric acid	H_2SO_4
Acetic acid	$HC_2H_3O_2$
Citric acid	$H_3C_6H_5O_7$

Here are the formulas for four of the bases we have used.

Sodium hydroxide   NaOH
Calcium hydroxide $Ca(OH)_2$
Potassium hydroxide   KOH
Ammonium hydroxide NH_4OH

Ammonium hydroxide is more correctly known as "aqueous ammonia." The term "ammonium hydroxide" is used here to emphasize that ammonia forms a base in water.

What do their formulas have in common?

Fig. 22 *Ammonia is the active ingredient in many household cleaners.*

Examine the following list of salts. How are their formulas different from those of acids and bases? How are they similar?

Sodium chloride   NaCl
Calcium carbonate $CaCO_3$
Calcium chloride $CaCl_2$
Sodium hydrogen carbonate $NaHCO_3$
Potassium chloride   KCl
Ammonium chloride NH_4Cl

Questions

1. (a) Suggest a way of classifying acids, bases, and salts using their chemical formulas.
   (b) Summarize any other ways of classification proposed by your classmates.

2. Use your method of classification to determine whether the following substances are acids, bases, or salts.
   (a) LiOH   (c) HBr   (e) $Al(OH)_3$ (g) KI
   (b) LiCl   (d) $BaSO_4$ (f) H_3PO_4 (h) NH_4NO_3

Chemically Speaking

NEUTRALIZATION

The product of the neutralization reaction between sodium hydroxide and hydrochloric acid is sodium chloride, or table salt. The term "salt" is now commonly used by chemists to describe any product that can be formed from a neutralization reaction. In general, a neutralization reaction always produces a "salt" and water.

$$acid + base \rightarrow \text{``salt''} + water$$

The word equation for this neutralization of hydrochloric acid by sodium hydroxide is:

hydrochloric acid + sodium hydroxide →

sodium chloride + water

The chemical equation for this neutralization of hydrochloric acid by sodium hydroxide is:

$$HCl + NaOH \rightarrow NaCl + H_2O$$

Notice that the "H" from the acid and the "OH" from the base give HOH or H_2O.

Investigation

9.12 Testing Household Products

The substances for this investigation include some widely used household products. Check the labels on some of these products and decide whether each one contains any acids, bases, or salts. If a label does not help you, predict whether the product is acidic, basic, or neutral. The following procedure will allow you to check your interpretations of labels and your predictions.

Materials

household solids: aspirin (ASA); baking soda (sodium hydrogen carbonate); cream of tartar; drain opener crystals; laundry detergent; toilet bowl cleaner crystals

household liquids or sprays: household ammonia; lemon juice; milk; oven cleaner; milk of magnesia; soft drink (colourless or pale in colour); shampoos

household cream or gel: hair remover

Choose some of these household products that you would like to test. Make a list of all the household acids and bases and the equipment that you will need. Give your list to your teacher for approval.

Be careful when drain opener crystals are added to water. The solution is corrosive and becomes very hot, and flammable gas is produced.

Some of these products are hazardous. Do not carry them to school. Your teacher will supply them.

Don't forget your safety goggles!

Method

1. Use the test methods you used in other investigations in this chapter to determine whether these substances are acidic, basic, or neutral.

2. You need only enough material to cover the bottom of a clean, rinsed test tube. In the case of a cream or gel, use a clean, dry glass rod to transfer enough material to cover the bottom of a test tube. Since indicators only work in water, you will have to add about 4 mL of water to each sample and shake to mix.

3. List all the steps in your investigation and let your teacher check them before you begin. Create a chart similar to those you used in other investigations to record your results or data.

Follow-up

1. Are there any surprises in your observations? If so, what are they?

2. What other acids and bases do you use in your daily life?

3. What role do acids and bases play in products you have examined? (You may need to consult reference materials.)

4. Select one common acid or base. Research how it was produced in ancient times and how it is produced today.

5. Write a brief report on the acid or base you chose in question 4. Share the information you have gathered with your classmates.

9.13 Acids and Bases in Action

Study Figures 23 to 31. Consider the numerous uses of acids and bases in daily life and in the manufacture of industrial and consumer products. The world would not be the same without acids and bases.

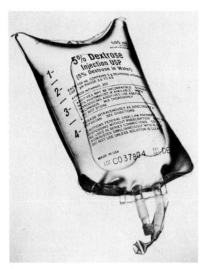

Fig. 24 *Nitric acid, HNO_3, is used in the manufacture of explosives, such as TNT. It is also the active ingredient for etching plates.*

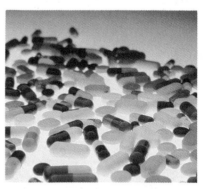

Fig. 25 *Sulfuric acid, H_2SO_4, is a most important chemical—being used to make dyes, drugs, detergents, and explosives. It is an ingredient in car batteries.*

Fig. 23 *Hydrochloric acid, HCl, is used in the manufacture of many drugs and for cleaning mortar from bricks. The manufacture of vinyl plastics also uses hydrochloric acids.*

Fig. 26 *Phosphoric acid, H_3PO_4, is an ingredient in cola drinks and rust removers, and is used in the manufacture of fertilizers.*

Fig. 27 *Ammonium hydroxide, NH_4OH, is the household ammonia used for various cleaning purposes.*

Fig. 28 *Sodium hydroxide, NaOH, commonly known as caustic soda, is used in the manufacture of soap, rayon, and paper.*

Fig. 29 *Calcium hydroxide, $Ca(OH)_2$, is the slaked lime used to make mortar and plaster for buildings.*

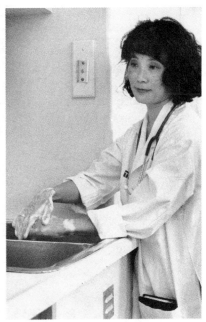

Fig. 30 *Potassium hydroxide, KOH, commonly known as caustic potash, is used to make soft soap. It is also the active ingredient in oven cleaners.*

Fig. 31 *Aluminum hydroxide, $Al(OH)_3$, is used in the manufacture of glass.*

P O I N T S · T O · R E C A L L

- Many acids and bases are poisonous and corrosive. They must be stored and handled carefully.
- Indicators change colour when they are in contact with an acid or a base. For example, acids turn blue litmus paper red, and bases turn red litmus paper blue.
- Phenolphthalein is colourless in an acidic solution and red in a basic solution.
- Acids and bases may change the appearance and texture of certain foods and may attack the surfaces of some food containers. Be careful when cooking food or choosing a food container.

- Acids react with metals to produce hydrogen; acids also react with carbonates or hydrogen carbonates to produce carbon dioxide.
- An acid can be recognized by the ''H'' that appears first in its formula. A base can be recognized by the ''OH'' in its formula.
- When acids react with bases, the bases are neutralized. Bases will also neutralize acids.
- Salts are solid compounds in crystal form. Neutral salts have no effect on litmus paper and phenolphthalein indicators.
- Solutions of acids, bases, and salts conduct electricity.

R E V I E W · Q U E S T I O N S

1. Why is it necessary to protect your hair and skin when using an oven cleaner spray?
2. Why should a container of hydrochloric acid never be stored on a high shelf?
3. When a concentrated acid solution is diluted with water, a considerable amount of heat is given off. Which would be safer, diluting the acid by adding a small amount of the acid to a lot of water, or by adding water to the acid? Explain.
4. Containers of many household cleaning products have hazardous warning labels. Why do you think they these labels are necessary?
5. Why should you never mix toilet bowl cleaner with a hypochlorite bleach?
6. Historians and scientists believe that many of the nobility in the ancient Roman Empire suffered from lead poisoning.
   (a) How did most of these people get the lead into their bodies?
   (b) Why did the poorer people not suffer in the same way?
   (c) Are there any ways we can get lead poisoning today?
7. List at least five household products for each category.
   (a) acids   (b) bases   (c) salts
8. What is an acid–base indicator?
9. Ceramic-tiled floors are often found in bathrooms and in kitchens. Why should the grout used between the tiles be acid resistant?
10. Why is vinegar added to beet salads?
11. Describe a chemical test to detect the presence of each of the following.
    (a) carbon dioxide gas
    (b) hydrogen gas
12. Given the formula of each chemical, predict whether it is an acid, base, or a neutral salt.

(a) H_2SO_3 (f) KOH
(b) $Mn(OH)_2$ (g) $HClO$
(c) HNO_2 (h) H_2S
(d) Na_2SO_4 (i) $Sr(OH)_2$
(e) $KClO_4$ (j) $BaBr_2$

13. In kettles, coffee makers, dishwashers, and steam irons, all of which use tap water, a scale will eventually build up on their inside surfaces. This scale is made up of materials such as calcium carbonate and magnesium carbonate.
    (a) How would you remove this scale safely?
    (b) Institutions such as hotels can run into expensive repairs if their water pipes become heavily scaled. What can they do to prevent scale from forming?
14. Marble tabletops, kitchen pastry boards, and bathroom vanities have been popular for many years. What precautions would you take when using them in your home?
15. Oxalic acid is sometimes used to remove rust stains. Why should you never use this acid to remove such stains from marble surfaces?
16. Pearls consist mostly of calcium carbonate crystals. Why should you not splash vinegar, orange juice, or wine on pearls?
17. When using an indicator to determine whether a solution is acidic or basic, why should you never use more than a few drops of the indicator?
18. Soap is slightly basic. If you use soap to wash hair or wool, why might you use vinegar in the final rinse water? Why are many commercial hair conditioners mildly acidic?
19. If you had spilled an acidic solution in the kitchen, how could you safely neutralize it?
20. A farmer may add slaked lime (calcium hydroxide) to the soil, because some crops grow poorly in soil that is too acidic. Describe what slaked lime does to the soil.

10 THE NATURE OF ACIDS, BASES, AND SALTS

Aluminum oxide ore is a valuable resource.

CONTENTS

10.1 Chemical Jeopardy

Let's play a game of "Chemical Jeopardy"! According to the rules of this game, the *answers* are provided but you must make up the *questions*. Our subject is water. For how many of these answers can you construct the corresponding questions?

100 points The answer is: Two hydrogen atoms and one oxygen atom.
What is the question?

200 points The answer is: Oxygen needs only two more electrons to complete its outermost occupied energy level.
What is the question?

300 points The answer is: H — $\ddot{\text{O}}$ — H
What is the question?

400 points The answer is: The decomposition of water into hydrogen gas and oxygen gas.
What is the question?

500 points The answer is: When you shut it off, the decomposition of water stops.
What is the question?

Playing "Chemical Jeopardy" helps you to think "in reverse." Sometimes, thinking in this way is a valuable skill in science, since it can help you to understand why things happen as they do. For example, you may have observed in investigation 5.1 that the electrolysis of water stopped as soon as the electrical current was cut off. Let's try again to work back from an answer to a question. The answer: Electrical energy can break down substances in electrolysis. The question: What sort of work can electrical energy do in a chemical reaction?

This question suggests another: How does electrical energy break down substances? In small groups, suggest an answer. Present your hypothesis to the class in the form of a diagram or a picture.

The next step after making a hypothesis is to test it. Can you explain the following on the basis of your hypothesis?

Although aluminum is the third most abundant element in the earth's crust, it was once considered a precious metal, like gold, because removing the aluminum from the aluminum oxide ore was very expensive. In the 1855 Exposition in Paris, the first aluminum ever produced was displayed alongside the Crown Jewels!

Nowadays, aluminum is removed from its ore by means of electrolysis. Since the province of Québec is rich in hydro-electric power, it is one of the major producers of aluminum in the world.

You will recall that energy in its many forms—electrical, solar, heat, mechanical, and so on—is the ability to do work.

10.2 Ion Formation

We learned in section 3.13 that an atom is electrically neutral because, in an atom, the number of electrons is equal to the number of protons. If an atom gains or loses electrons, it becomes charged and is then called an **ion**.

Since noble gases are very unreactive, chemists believe that there is something special about the arrangement of electrons in noble gas atoms. But how are the structures of noble gas atoms related to the structures of ions?

A neutral atom of sodium has one electron in its outermost occupied energy level. If that electron is lost, the electronic structure of the resulting sodium ion resembles the structure of a neon atom (Figure 2a). The sodium ion has 11 protons and 10 electrons. Therefore, the ion's net charge equals the charge on one proton. We call this net charge a "positive one charge" and give it the symbol 1+.

Magnesium also forms a positive ion. In order to resemble an atom of the noble gas neon, a magnesium atom loses two electrons from its outermost occupied energy level (Figure 2b). The magnesium ion therefore has a net charge equal to the charge on two protons. We represent this net charge as 2+.

Fig. 1 *Salt, used in many foods, contains positively and negatively charged ions.*

Atoms that have lost electrons have more protons than electrons. These atoms have become *positively* charged ions. Atoms that have gained electrons have more electrons than protons. These atoms have become *negatively* charged ions.

What about nonmetals? When an oxygen atom gains two extra electrons, it becomes an ion with a negative two charge, 2− (Figure 2c). A fluorine atom can achieve an octet of electrons in its valence shell by gaining one electron to give an ion with a negative one charge, 1− (Figure 2d). The ions from both oxygen and fluorine atoms are similar to a neon atom in electronic structure.

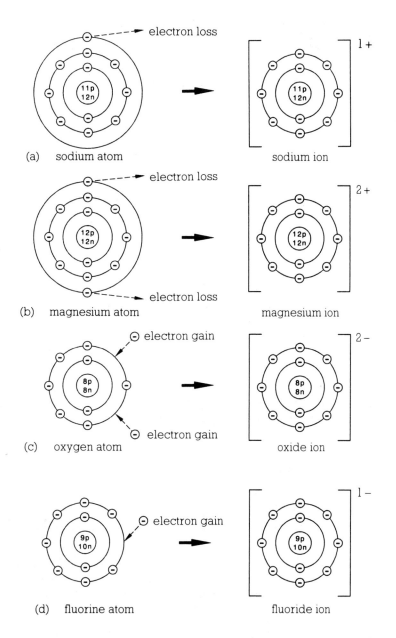

Fig. 2 *The ions formed from sodium, magnesium, oxygen, and fluorine are similar to the atom of the noble gas neon in electronic structure.*

Recall that the electrons in the outermost occupied energy level of an atom are called valence electrons. The outermost occupied energy level is called the valence shell. In the examples above, we see that only the valence electrons of the atoms take part in ion formation.

We have represented atoms and ions by means of Bohr-Rutherford diagrams. You may have also learned about Lewis diagrams. They are a simpler method for representing the valence electrons of atoms than are Bohr-Rutherford diagrams (Figure 3). If you have yet to study Lewis diagrams or need a review, refer to section 5.5.

In a Lewis diagram we simply represent an atom by its symbol and a set of dots around the symbol to indicate the valence electrons of that atom.

Fig. 3 *The Lewis diagram is much simpler than the Bohr–Rutherford diagram.*

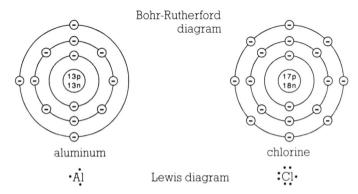

Bohr-Rutherford diagram

aluminum   chlorine

·Al   Lewis diagram   :Cl·

Questions

1. The number of valence electrons in a lithium atom is 1; the number in a fluorine atom is 7. Look at the elements in Groups 1 to 8 of the periodic table. How is the group number related to the number of valence electrons in an atom of an element in that group?

2. If a sodium atom loses its only valence electron, the atom becomes a sodium ion. Draw the Lewis diagram of a sodium ion.

3. If a fluorine atom takes an extra electron into its valence shell, it becomes a fluoride ion. Draw the Lewis diagram of this fluoride ion.

10.3 Ionic Bonding

We have just learned that, when a metal atom has lost all its valence electrons, its electronic structure is similar to that of the nearest noble gas atom. A nonmetal atom can gain one or more electrons to achieve an electronic structure similar to that of the nearest noble gas atom. Because there are eight electrons in the valence shell of a noble gas atom (except helium), this idea became known as the **octet rule**. Now you know why eight is a "magic number"!

Remember that "oct" means eight. Octet means a group of eight.

Using a Lewis diagram to represent an ion then becomes a much easier task. With a metal atom, we simply remove all the valence electrons. The ion is then represented by a symbol and a charge.

· Mg ·   loses two electrons and becomes $[Mg]^{2+}$

Al ·   loses three electrons and becomes $[Al]^{3+}$

With a nonmetal atom, we add sufficient electrons to make a total of eight. The ion is then represented by a symbol with eight dots around it, plus the charge.

:S: gains two extra electrons to become $[:S:]^{2-}$

:P·   gains three electrons to become $[:P:]^{3-}$

The common exception to the rule is the nonmetal hydrogen. A hydrogen atom may gain one electron to form an ion that resembles an atom of the noble gas helium. This noble gas atom is stable with just two valence electrons.

H·   gains one extra electron to become $[·H·]^{1-}$

In some cases, a hydrogen atom can also lose its valence electron, and becomes a positive ion, $[H]^{1+}$.

Ion formation is used to explain how a metal and a nonmetal join (or bond) together to form a compound. A main group metal atom loses all its valence electrons to become a positive ion. This loss usually exposes a complete inner octet. (An exception occurs when the ion has the same number of electrons as a helium atom.) The nonmetal atom gains sufficient electrons to complete an octet in its valence shell. The nonmetal atom thus becomes a negative ion. The positive and negative ions attract each other, and this attraction is called an **ionic bond**.

It is the strong ionic bond in aluminum oxide that must be broken by means of electrical energy in order to obtain the pure aluminum.

Because ionic compounds have very high melting points, they are usually solid at room temperature. Most ionic compounds dissolve at least a little in water.

Let's see how the compound lithium fluoride is made from the elements lithium and fluorine. Locate lithium and fluorine in the periodic table, and using their group numbers as a guide, draw the Lewis diagrams for the atoms of these two elements in your notebook.

How many electrons must the metal lithium lose in order to resemble the noble gas helium? How many must the nonmetal fluorine gain in order to resemble the noble gas neon? How do you think these two elements interact in order to give each of the elements the "noble gas look"? Illustrate this interaction in your notebook.

The compound produced is called lithium fluoride. What is the charge on the lithium ion? What is the charge on the fluoride ion?

We have learned that matter is electrically neutral. Can you explain how an ionic compound can be neutral if it is composed of positive and negative ions? How does the fact that matter is neutral help you to check if the chemical formula of an ionic compound is correct?

Some atoms lose or gain more than one electron. Follow the same procedure that you used above to show the formation of sodium oxide from atoms of sodium and oxygen. How many sodium atoms are needed to react with each oxygen atom? What is the formula of sodium oxide?

Now you are ready for a real challenge. Using Lewis diagrams, show the formation of magnesium nitride from atoms of magnesium and nitrogen. Check that the ionic charges add up to 0. What is the formula of magnesium nitride?

When we write the formula of the compound, we must leave out the ionic charges. For example, the formula of potassium bromide is KBr, not $K^{1+}Br^{1-}$.

Questions

1. Use a Lewis diagram to represent the ion formed from each of the following atoms:

   (a) N   (e) Cl   (h) Br
   (b) O   (f) K   (i) Li
   (c) Na   (g) Ca   (j) Ga
   (d) Be

Fig. 4 *Iron oxide (rust) is a binary ionic compound.*

2. (a) Use a Lewis diagram to show the ionic bonding between each of the following pairs of elements:

(i) Li and Cl	(xi) Ca and O
(ii) Li and S	(xii) Al and O
(iii) Al and Cl	(xiii) Be and F
(iv) Be and Br	(xiv) Mg and P
(v) Sn and As	(xv) Sr and F
(vi) K and Br	(xvi) Na and S
(vii) Mg and Cl	(xvii) Ca and P
(viii) Cs and P	(xviii) Ba and Cl
(ix) Rb and F	(xix) Al and N
(x) Al and I	(xx) Ba and N

(b) State the formula for each compound in (a).

3. In section 5.5, question 1, you drew Lewis diagrams for six binary compounds. Three of them were compounds of a metal and a nonmetal (NaCl, KI, and MgO). In those three Lewis diagrams, you represented the bond between the metal and the nonmetal by means of shared electrons. In the present section, you have assumed that the electrons are transferred from a metal to a nonmetal when ions are formed.

(a) For each compound (NaCl, KI, and MgO), draw a Lewis diagram that illustrates the sharing of electrons and another one that illustrates electron transfer.

(b) Is there any basis for deciding which type of Lewis diagram is more reasonable? If so, explain. If not, what type of evidence do you need to make a decision?

10.4 A Second Look at Writing Chemical Formulas

Chemists obviously do not draw a Lewis diagram each time they want the chemical formula of a compound that contains a metal and a nonmetal. Chemists look for a shortcut. You already derived such a shortcut using valences and the crossover rule in section 5.4.

Recall that the term "valence" means the number of bonds an atom can form.

Questions

1. You can look up the valences of the main group elements from the periodic table. How is the group number related to the valence of the elements in each group from 1 to 8?

2. Review the crossover rule and then give the correct formula for each of the following:

(a) potassium sulfide
(b) barium nitride
(c) aluminum oxide
(d) magnesium bromide
(e) radium chloride

(f) calcium hydride
(g) lithium iodide
(h) sodium phosphide
(i) beryllium sulfide
(j) strontium hydride

3. As compounds are electrically neutral, the charges on any ions present must "balance".

(a) Check each formula you derived in question 2 to make sure that each compound is electrically neutral.

(b) From your examination of the formulas in (a), can you propose an alternative to the crossover rule for determining the formula of a compound that contains a metal and a nonmetal?

10.5 Covalent Bonding

In the last few sections, we studied a model of how a metal atom and a nonmetal atom might be held together by an ionic bond. In this model, the metal atom donates its valence electrons to the nonmetal atom, and both atoms become ions that resemble a noble gas in electronic structure.

Now consider a binary compound of two nonmetals. If nonmetal atoms reach the electronic structure of noble gas atoms by gaining electrons, how can two nonmetal atoms gain electrons at the same time? The answer lies in the idea of sharing. The nonmetal atoms can both gain electrons at the same time only by *sharing* their valence electrons. This sharing of electrons is called a **covalent bond**. Covalently bonded compounds usually have lower melting points than ionically bonded compounds. Many covalently bonded compounds are liquids or gases at room temperature. Only a few of them dissolve in water.

Why does covalent bonding hold a molecule together? As with ionic bonding, the answer lies in electrical attraction and repulsion. Shared electrons, which are negative, simultaneously attract both atomic nuclei, which are positive. Also, the presence of the negative electrons between the nuclei reduces the repulsion between the nuclei.

Now answer the following questions on the assumption that our present models of bonding are correct. In other words, assume that:

- A metal atom and a nonmetal atom bond by means of electron transfer and the attraction of oppositely charged ions.
- Two nonmetal atoms bond by means of electron sharing and the attraction of both positive nuclei to the shared negative electrons.

Questions

1. Indicate the type of bond — ionic or covalent — that forms between each of the following pairs of atoms:
   (a) H and Cl
   (b) C and S
   (c) Sr and Br
   (d) Ca and F
   (e) K and S
   (f) Si and O
   (g) K and N
   (h) Cr and O
   (i) S and O
   (j) Mg and Cl
   (k) P and O
   (l) O and O
   (m) Li and O
   (n) Fe and Cl
   (o) Cu and I
   (p) C and Cl
   (q) N and O
   (r) I and Cl
   (s) Al and N
   (t) Ba and P

2. A molecule of ammonia contains one atom of nitrogen and three atoms of hydrogen. Draw the Lewis diagram to represent this molecule.

3. The chemical formula of aluminum fluoride, AlF_3, resembles the formula for ammonia, NH_3.
   (a) Draw a Lewis diagram to represent aluminum fluoride.
   (b) Compare your answer for (a) to your answer for question 2. Are we justified in referring to both diagrams as models of "molecules"? Explain.

4. Methane is the chemical name for the natural gas used in gas stoves and home-heating furnaces. A methane molecule contains one atom of carbon and four atoms of hydrogen. Draw the Lewis diagram to represent a molecule of methane.

5. Review your rules for naming compounds that contain only nonmetals, and then draw Lewis diagrams to represent molecules of each of the following compounds:
   (a) hydrogen fluoride
   (b) sulfur dibromide
   (c) oxygen difluoride
   (d) carbon monoxide
   (e) nitrogen trifluoride
   (f) carbon tetrachloride
   (g) carbon dioxide
   (h) hydrogen sulfide

Covalent compounds of hydrogen are not named systematically. If they were, "hydrogen fluoride" would be "hydrogen monofluoride" and "hydrogen sulfide" would be "dihydrogen monosulfide."

10.6 What's so Radical About Polyatomic Ions?

A **radical** is a group of atoms of different elements — usually nonmetals — bonded together to form a single ion. Modern chem-

Compounds that contain polyatomic ions must contain at least three elements. Compounds of three elements are called **ternary** compounds. Those with four elements are **quaternary** compounds.

Remember that compounds are electrically neutral.

ists refer to such an ion as a **polyatomic ion**. In writing formulas, you treat the polyatomic ion as a "package deal," as if it were an ion formed from a single atom. The sulfate ion consists of one sulfur atom and four oxygen atoms (SO_4) and behaves as if it were an element with a valence of 2. The valences of many polyatomic ions can be found at the end of this book. Most common polyatomic ions are negatively charged, like the sulfate ion, SO_4^{2-}. The ammonium ion, NH_4^{1+}, is an exception.

The formula of sodium sulfate is Na_2SO_4. Make sure you understand why this formula is correct, given the valences of the sodium and sulfate ions. Another way of looking at this is to make sure that the charges on two sodium ions balance the charge on a sulfate ion.

What is the formula of the compound that contains sodium ions and phosphate ions? magnesium ions and phosphate ions?

You have already learned that the crossover rule is consistent with the balancing of ionic charges for binary ionic compounds. The crossover rule can also be used for predicting the formulas of compounds that contain polyatomic ions.

	Lithium nitrate	Aluminum sulfate	Magnesium carbonate
Step 1 Write down the symbols in the order given in the name	Li NO_3	Al SO_4	Mg CO_3
Step 2 Record the valence value for each element or radical given	1 1 Li NO_3	3 2 Al SO_4	2 2 Mg CO_3
Step 3 Crossover the valence values	$Li_1(NO_3)_1$	$Al_2(SO_4)_3$	$Mg_2(CO_3)_2$
Step 4 Find the highest factor common to the two valence values	1	1	2
Step 5 Divide the two valence values by this highest factor	$Li_1(NO_3)_1$	$Al_2(SO_4)_3$	$Mg_1(CO_3)_1$
Step 6 Drop any "1" in the formula	$LiNO_3$	$Al_2(SO_4)_3$	$MgCO_3$

In step 3, the brackets indicate that the whole polyatomic ion is a single group. The "3" in NO_3, the "4" in SO_4, and the "3" in CO_3 are not part of the crossover rule, and therefore are not divided or dropped in steps 5 and 6.

The brackets are kept if, after you have applied the crossover rule, there are two or more polyatomic ions in the formula.

If you are given the formula for a compound that contains a polyatomic ion, you need not look up the ion in a table to figure out its valence and charge. All you need is a little "Chem Smarts."

Let's begin with the compound sodium cyanide, NaCN. Its formula shows one Na and one CN. Consider also the compound calcium carbonate, $CaCO_3$. Its formula shows one Ca and one CO_3.

Now here is where thinking "in reverse" comes in handy! All you have to do to find the valence and charge of each polyatomic ion is to apply the crossover rule in reverse! Just follow these steps in the chart:

	Sodium cyanide	Calcium carbonate
Step 1 Write down the formula of the compound.	Na CN	Ca CO_3
Step 2 Record the valence values that had been crossed over in writing this formula.	Na_1 CN_1	Ca_1 $(CO_3)_1$
Step 3 Reverse the crossover of the valence values.	1 1 Na CN	1 1 Ca CO_3
Step 4 Is the valence value correct for the ion you recognize? If not, what should it be?	1 YES Na	2 NO Ca
Step 5 If it is *not correct*, then multiply both values by the corrected valence value.	1 1 Na CN	2 2 Ca CO_3
Step 6 Write the valence values as charges. The charge on the first ion is positive; the second is negative.	Na^{1+} CN^{1-}	Ca^{2+} CO_3^{2-}

As a result of this exercise, we have determined that the cyanide ion has a valence of 1 and a charge of 1−, and that the carbonate ion has a valence of 2 and a charge of 2−.

Remember that many polyatomic ions, such as CO_3^{2-} and NO_3^{1-}, contain subscript numbers in their formulas. Be careful not to cross these over. Assume there is only one polyatomic ion in a formula unless it is enclosed in brackets.

By now, you may be comfortable with the balancing of ionic charges. You may not need to write out all the above steps to determine the charge on a polyatomic ion. To find out, try your own method for determining the charge on the cyanate ion in calcium cyanate, $Ca(CNO)_2$, and the arsenite ion in sodium arsenite, Na_3AsO_3. If you obtain answers of 1– and 3–, respectively, your own method is working well. The following questions will give you additional practice.

Questions

1. (a) Using the reverse crossover method explained above, determine the valence of and charge on a sulfite ion, SO_3, given that the compound calcium sulfite has the formula $CaSO_3$.
   (b) What is the formula of each of the following?
      (i) sodium sulfite
      (ii) aluminum sulfite
      (iii) ammonium sulfite

2. By whatever method you choose, determine the charge on the following underlined polyatomic ions.
   (a) $NaHSO_4$ (d) $K_2Cr_2O_7$ (g) $BaCrO_4$
   (b) $Ca(ClO_4)_2$ (e) $KMnO_4$ (h) $Al(ClO)_3$
   (c) $Na_2S_2O_3$ (f) $LiBH_4$ (i) Na_4SiO_4

3. Use the symbols and valence values of the elements and polyatomic ions given at the back of this textbook to write the correct formula for each of the following:
   (a) sodium acetate
   (b) strontium hypochlorite
   (c) lithium iodate
   (d) zinc phosphate
   (e) calcium chlorite
   (f) sodium dihydrogen phosphate
   (g) potassium hydrogen sulfite
   (h) potassium hydrogen sulfide
   (i) aluminum phosphate
   (j) barium nitrate
   (k) strontium dichromate
   (l) aluminum hydrogen carbonate
   (m) ammonium bromate
   (n) aluminum carbonate
   (o) magnesium hydroxide
   (p) silver nitrate
   (q) sodium perchlorate
   (r) cadmium nitrite
   (s) zinc chromate
   (t) aluminum sulfite
   (u) sodium chlorate
   (v) potassium nitrate

4. You have seen that a compound ending with *-ate* or *-ite* contains a polyatomic ion. (Look at the names of the compounds in question 3.) Exceptions to this naming rule are compounds containing the ammonium ion, NH_4^{1+}, the hydroxide ion, OH^{1-}, and the cyanide ion, CN^{1-}. Write the correct formula for each of the following:

   (a) calcium hydroxide
   (b) magnesium cyanide
   (c) ammonium sulfide
   (d) ammonium hydroxide

The *-ide* ending in hydroxide and cyanide does not mean that the compound is binary.

5. In completing question 3, you referred to a table of valence values. Most of the compounds listed in question 3 contain a metal and a polyatomic ion.

   (a) What alternative method could you use to determine the valence values of the metals?
   (b) Are there any metals in question 3 for which your alternative method does not work? Explain.

Investigation

In an electrolysis cell the current flows from the cathode (the negative electrode) to the anode (the positive electrode). In other words, electrons flow from the anode to the cathode.

10.7 Electrolyte or Nonelectrolyte?

An **electrolyte** is a compound that conducts electricity when it is dissolved in water. A **nonelectrolyte** does not conduct electricity when dissolved in water. Do you remember any investigation where it was necessary to add an electrolyte to water in order for the current to flow? Do you remember any compounds you have studied that are electrolytes? Is there a way to tell in advance whether a compound will be an electrolyte? Discuss the answers to these questions with your group. If you think that it is possible to predict which of the test substances are electrolytes, make your predictions and discuss them with your group.

To determine whether a compound is an electrolyte or not, you will have to dissolve it in water and then test its solution for conductivity.

Don't forget your safety goggles!

Do not test solid sodium hydroxide and pure liquid sulfuric acid. These chemicals are highly corrosive.

Handle pure liquid acetic acid with care and keep it in the fume hood.

Materials

Your teacher will provide you with samples of pure compounds, as well as their solutions. Possible choices include sugar ($C_{12}H_{22}O_{11}$), vinegar (acetic acid, $HC_2H_3O_2$), ethyl alcohol (C_2H_6O), sulfuric acid (H_2SO_4), sodium hydroxide (NaOH), and calcium chloride ($CaCl_2$). Be sure to write down the name and formula for each substance you test.

Make a list of the equipment you will need to run these tests and ask your teacher to approve the list.

Handle solutions of acids and
bases with care.

Method

1. Thinking of the procedure you developed in investigation 9.9,
   plan how you will conduct this test. Get your teacher's approval
   before beginning.

2. Prepare a data chart, similar to the one below, for recording
   your observations. Take special note of whether the substance is
   a good conductor or a poor one.

Substance	Formula	Conductivity of pure substance	Conductivity of solution		
			good?	poor?	none?

Follow-up

1. Did any of the pure compounds you tested conduct electricity?

2. (a) Which compounds are electrolytes?
   (b) Which compounds are nonelectrolytes?

3. Were all the electrolytes equally good at conducting electricity?

4. The electrolytes that carried enough current to light the bulb
   brightly are called **strong electrolytes**. The ones that carried
   only enough current to make the bulb glow faintly are **weak
   electrolytes**. Divide your list of electrolytes into strong and
   weak ones.

5. Prepare a report summarizing your observations in this
   investigation.

Investigation	## 10.8 Explaining Conductivity

Put on your thinking caps and break up into groups. In your
groups, discuss the following question: Why do some compounds
conduct electricity in solution while others do not? Base your
discussion on the lists of electrolytes and nonelectrolytes that you
made in investigation 10.7, question 2. (The formulas of the
compounds you tested should help.) Your goal is to come up with
a hypothesis to explain conductivity. In other words, you must

decide what happens in a solution of an electrolyte that allows it to complete an electrical circuit. Draw a picture to illustrate your hypothesis.

Compare your group's hypothesis with those from other groups. Now is the time to test the different hypotheses!

Use your hypothesis to predict whether solutions of the following compounds will be electrolytes or nonelectrolytes: hydrochloric acid (HCl), acetic acid ($HC_2H_3O_2$), sodium hydroxide (NaOH), sodium chloride (NaCl), calcium hydroxide ($Ca(OH)_2$), and copper(I) chloride (CuCl). (You may have already tested some of them.)

Solutions of copper compounds are toxic.

Materials

test solutions listed above
conductivity apparatus from investigation 10.7

Method

Use the same method as for investigation 10.7. (Make sure that you answer follow-up question 2 before you put away your apparatus.)

Follow-up

1. What experimental observation tells you that a substance is a good conductor of electricity?

2. Dilute one of the solutions that is a good conductor by adding a small volume of it to a larger volume of water.
   (a) Is the conducting ability of the solution affected by dilution?
   (b) What condition must you keep constant in order to make a fair comparison of the conducting ability of different substances?

3. Explain how strong electrolytes are able to conduct electricity in solution.

4. You learned in investigation 10.7 that strong electrolytes are good conductors in solution, whereas weak electrolytes are only fair conductors in solution.
   (a) Does your hypothesis for explaining conductivity account for this difference in behaviour?
   (b) If not, how can you modify your hypothesis to account for the different conducting abilities of strong and weak electrolytes in solution?

5. In section 10.3, question 3, you compared two types of Lewis diagrams for binary compounds of a metal and a nonmetal. Do your observations in the present investigation help you decide which Lewis diagram is more appropriate for each compound? Explain your answer.

10.9 Ions and Technology

Many technological applications depend on the existence of ions. This section indicates just a few of them.

Fig. 5 *(a) The operation of a car battery depends on ions in solution.*

Fig. 5 *(b) Some people treat water with an ion exchange resin before pouring the water into an iron.*

Fig. 5 *(c) Laundry softeners and anti-static sprays reduce ion build-up on fabrics.*

CHEMICAL TIDBITS

DO IONS AFFECT PEOPLE?

Different atmospheric conditions can lead to a buildup of positive or negative ions in the air. On humid days there tend to be more positive ions, and some people think that this is why some of us feel down. Clinical studies have also shown that with a buildup of positive ions in the air, some people have more headaches, others feel more fatigued, and others suffer more from sinus and asthma conditions. The air immediately after a thunderstorm tends to contain more negative ions. Some people have said that they feel better after a storm. Today you can actually buy a negative ion generator for your home (Figure 6).

Fig. 6 *Can commercially produced negative ion generators improve your health and mood?*

CHEMICAL TIDBITS

ELECTROPLATING

Electroplating uses an external source of electrical energy to coat an object with a particular metal (Figure 7). The most common metals used for plating are chromium, nickel, cadmium, tin, silver, copper, and gold.

All electroplating is done in essentially the same way. The electrolyte used must be an ionic compound containing the metal ion to be used for plating. The anode is also made from the metal that is being used for plating. The cathode is the object to be plated. This object must conduct electricity and is therefore usually metallic.

Copper plating can be used as an example. The electrolyte solution is a mixture containing copper(II) sulfate. The anode is a clean piece of copper; the cathode is a metal object to be plated. As the current flows through the solution, a layer of copper builds up on this object.

Successful electroplating depends on the cleanliness of the object being plated, the purity of the electrolyte solution, the strength of the electric current, and the time taken.

Many toxic chemical are used in the cleaning and plating processes of electroplating. The wastes from electroplating plants must be carefully handled to avoid polluting the environment.

Fig. 7 *Electroplating is used to produce an attractive and durable finish on many products.*

CHEMICAL TIDBITS

A porous material has tiny holes in it that let gases and liquids pass through.

Steps in anodizing aluminum:

preparation of the object

↓

water rinse

↓

NaOH etching of surface

↓

water rinse

↓

acid cleaning

↓

water rinse

↓

anodizing

↓

water rinse

↓

dyeing

↓

water rinse

↓

sealing

ANODIZING OF ALUMINUM

Aluminum rapidly forms an oxide layer which sticks to the surface so strongly that further corrosion of the metal cannot take place. It is therefore an ideal metal to use in places where it will be exposed to the environment. However, with time, the oxide layer becomes very dull and unsightly.

Anodizing the aluminum is an electrical process that prepares the surface of the aluminum so that coloured dyes can be applied. The aluminum surface is first coated with a layer of porous oxide in an electrical process. The oxide layer can accept dye, which is then sealed to the surface by means of boiling. The result is a better looking and longer lasting product. Anodized aluminum is frequently used in household products, including aluminum siding and jewellery (Figure 8).

Fig. 8 *Jewellers use the anodizing process to produce beautifully coloured aluminum jewellery.*

Questions

1. What types of consumer products can you name that produce or exchange ions? How are these devices used in everyday life?

2. Explain, in your own words, the processes of electroplating and anodizing.

3. Why is it dangerous to have an electrical device, such as a radio or a hair dryer, nearby when you are taking a bath?

4. The human body is made up of more than 90% water. In this water are dissolved compounds, some of them ionic, that our bodies need to function well. With this in mind, why is it safer to be in an automobile rather than outside during a thunderstorm?

10.10 Testing Salts With Litmus

Ammonium compounds have been used as "smelling salts" to revive people who have fainted.

Sodium carbonate can be bought in a supermarket as washing soda. Boxes of washing soda are labelled "corrosive," as are containers of lye, which is sodium hydroxide. Again, like lye, washing soda can be used to clear drains. Would you have thought that sodium carbonate, a salt, could be corrosive enough to clear a clogged drain?

In this investigation we will test several salts with litmus. What do you think will happen?

You probably remember that sodium hydrogen carbonate is popularly known as baking soda.

You have considerable experience now in detecting acids and bases with indicators. Put your knowledge to work in testing salts like ammonium chloride (NH_4Cl), sodium carbonate (Na_2CO_3), sodium hydrogen carbonate ($NaHCO_3$), and aluminum sulfate ($Al_2(SO_4)_3$).

Aluminum sulfate is used in making paper and turning animal skins into leather.

Materials

Draw up your own materials list and have your teacher approve it.

Don't forget to wear your safety goggles!

Method

1. Recall how you tested other substances for acidic or basic properties with litmus paper, and set up a similar test for the substances in your list.

2. Get your teacher's approval of your testing procedure before you begin.

3. Put your findings in a data chart.

Sodium carbonate is quite corrosive. Handle it with care.

Follow-up

1. (a) What did you predict would happen when you tested the salts with litmus paper?
   (b) Were you surprised by the results of the test?

2. Were any of the salts you tested neutral?

3. Which salts acted like acids and which acted like bases?

4. Does your present definition of an acid or a base explain these observations?

5. You know that carbonates and hydrogen carbonates react with acids. Is this information consistent with your findings in the present investigation? Explain.

6. What must a scientist do when a theory can no longer explain an experimental observation?

10.11 Theories About Acids and Bases

It may have seemed curious to you, in investigation 10.10, that substances containing no hydrogen or hydroxide in their formulas could behave as if they do. This observation has puzzled chemists for many, many years.

Antoine Lavoisier (Figure 9) was a remarkable scientist whose careful and organized study of the element oxygen added a great deal to scientific knowledge. His studies of this element led him to believe that oxygen was an important component of all acids. The name "oxygen," which he gave to the element, means "acid-former."

While some acids such as sulfuric acid, H_2SO_4, contain oxygen, other acids such as hydrochloric acid, HCl, contain none at all. There must be more to acids than the presence of one or more oxygen atoms in the formula.

Svante Arrhenius (Figure 10) proposed that the presence of hydrogen in the formula was a better indicator of acidic behaviour. He defined an acid as a chemical that gives up a hydrogen ion in water. Both acetic acid and hydrochloric acid release their hydrogen ions in water (Figure 11).

Antoine Lavoisier said that only compounds that contain oxygen can be acids.

Svante Arrhenius said acids are compounds that give up H^{1+} ions in water.

Fig. 9 *(left) Antoine Lavoisier (1743–1794) believed that oxygen was an important component of all acids.*

Fig. 10 *(right) Svante August Arrhenius (1859–1927).*

Fig. 11 *Most acetic acid molecules (sometimes represented as HAc) do not break up into ions in water. Almost all hydrochloric acid molecules do break up into ions in water.*

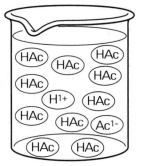 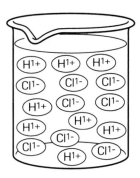

Bases give up OH^{1-} ions in water.

Arrhenius suggested furthermore that bases are compounds that give up hydroxide ions in water. The well-known bases sodium hydroxide, NaOH, and calcium hydroxide, $Ca(OH)_2$, certainly support this definition of a base (Figure 12).

The presence of ions in acidic and basic solutions explains why these solutions are conductors of electricity.

Fig. 12 *Both sodium hydroxide, NaOH, and calcium hydroxide, $Ca(OH)_2$, exist as ions in water.*

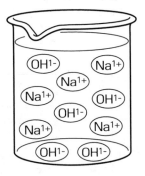 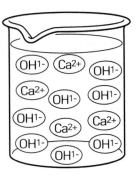

J. N. Brönsted and J. M. Lowry said acids are hydrogen ion donors.

The ion with the formula H_3O^{1+} is called the **hydronium ion**. Essentially, it is a water molecule with a hydrogen ion attached to it.

Bases are hydrogen ion acceptors. If a base accepts a hydrogen ion from water, a hydroxide ion is formed.

Unfortunately, the Arrhenius theory does not explain the behaviour of some salts. You have observed, for example, the acidic behaviour of ammonium chloride and the basic behaviour of sodium carbonate.

Simultaneously but separately, J. N. Brönsted in Denmark and J. M. Lowry in the United States came up with another way of defining acids and bases. Their definition took in non-neutral salts as well. They defined an **acid** as a substance that donates (gives away) a hydrogen ion to water. Both hydrochloric acid and ammonium chloride can do this.

Brönsted and Lowry define a base as a substance that accepts a hydrogen ion from water. Both the traditional base sodium hydroxide and the corrosive salt sodium carbonate can play the role of a base in water.

Questions

1. Which of these theories best explains the behavior of acids and bases?

2. Which theory can explain all the observations of acidic and basic behaviour you have met so far?

3. Hydrogen is an unusual element in that it can form a positive or a negative ion in chemical reactions, as well as forming covalent compounds.
   (a) What are the formulas and names for the two ions of hydrogen?
   (b) Which ion is thought to be responsible for acidic properties?

4. (a) Predict whether hydrogen and chlorine atoms in the substance hydrogen chloride, HCl, are held together by ionic or covalent bonds.
   (b) Hydrochloric acid is a solution of hydrogen chloride in water. Is the type of bonding between hydrogen and chlorine the same in hydrochloric acid as it is in pure hydrogen chloride? Explain.

P O I N T S · T O · R E C A L L

- Ions form when atoms lose or gain electrons.
- The electronic structure of an ion of a main group element resembles the electronic structure of the nearest noble gas atom in the periodic table.
- Lewis diagrams represent only the valence electrons of an atom or ion.
- Main group metals form positive ions by losing all their valence electrons.
- Nonmetals form negative ions by gaining sufficient electrons to obtain eight electrons in their valence shell.
- According to the octet rule, an atom loses or gains enough electrons so that there are eight electrons in the outermost occupied shell.
- Hydrogen is an exception to the octet rule. It needs only one more electron to fill its valence shell.
- Ionic bonding occurs when a positive ion is attracted to a negative ion.

- The crossover rule may be used to write the formula of an ionic compound.
- A ternary compound contains three elements.
- Compounds whose names end in *-ate* or *-ite* contain oxygen.
- The term ''valence'' refers to the number of bonds an atom of an element can form.
- A radical is a group of atoms that bond together to form a polyatomic ion.
- When two nonmetals combine to form a compound, they form covalent bonds. In a covalent bond, the atoms share electrons in order to obtain the electronic structure of a noble gas.
- Ionic compounds are usually solid at room temperature and are more likely to dissolve in water than are covalent compounds.
- An electrolyte is a substance that dissolves in water to form a conducting solution.
- Soluble acids, bases, and salts are electrolytes in water. All soluble salts and certain acids and

bases are strong electrolytes because all their molecules break up into ions in water. Other acids and bases are weak electrolytes because fewer of their molecules break up into ions in water.

- Unless they are acids or bases, covalently bonded compounds are nonelectrolytes. Their molecules do not break up into ions in water.
- Electroplating uses electrical energy to cover a metal object with a layer of another metal.

- In electroplating, the object to be plated is the cathode, and the electrolyte must contain ions of the plating metal.
- Certain salts, such as those containing the ammonium or aluminum ion, have acidic properties.
- Certain salts, such as those containing the carbonate, hydrogen carbonate, or acetate ion, have basic properties.

R E V I E W · Q U E S T I O N S

1. What is the relationship between ion formation and the electronic structure of the atom of the nearest noble gas?
2. What type of ion forms when a neutral atom gains an electron?
3. What type of ion forms when a neutral atom loses an electron?
4. What is a ternary compound?
5. In general, which type of compound—ionic or covalent—dissolves better in water?
6. What information is given when the name of a compound ends in -ide?
7. What information is given when the name of a compound ends in -ate or -ite?
8. Give an example of a positive polyatomic ion.
9. Give an example of a negative polyatomic ion.
10. Use a Lewis diagram to represent the ion formed from each of the following atoms:
    (a) P   (c) Mg   (e) N   (g) Cl
    (b) Na   (d) Al   (f) S   (h) K
11. Use a Lewis diagram to show the ionic bonding between each of the following pairs of elements:
    (a) Na and Cl   (d) Mg and S   (g) Al and P
    (b) K and N   (e) Ca and Br   (h) Li and O
    (c) Ca and N   (f) Cs and F   (i) Sr and P
12. Predict the type of bonding in each of the following compounds:
    (a) NH_3 (e) H_2O_2 (i) CO (m) P_2O_5
    (b) BaS (f) MnO_2 (j) AlN (n) $NaOH$
    (c) Li_3N (g) CaO (k) CdO
    (d) $CaCl_2$ (h) KH (l) Fe_2O_3

13. Predict whether the following compounds will be electrolytes in water.
    (a) H_2 (e) KCl (i) H_3PO_4
    (b) CH_4O (f) Na_3PO_4 (j) $Sr(OH)_2$
    (c) HNO_3 (g) $C_6H_{12}O_{11}$
    (d) NH_4Cl (h) $MgSO_4$
14. (a) If you wanted to plate a medal with silver, what electrolyte would you choose for the electroplating solution?
    (b) To which electrode would you attach the medal?
15. Predict whether the following salts will be acidic, basic or neutral.
    (a) $(NH_4)_2SO_4$ (c) K_2CO_3 (e) NaI
    (b) $Mg(C_2H_3O_2)_2$ (d) NH_4NO_3 (f) $SrSO_4$
16. Write the correct formula for each of the following compounds.
    (a) barium phosphate
    (b) calcium nitrate
    (c) ammonium sulfide
    (d) potassium hydroxide
    (e) aluminum nitrate
    (f) beryllium sulfate
    (g) lithium carbonate
    (h) magnesium hydrogen carbonate
    (i) strontium phosphate
    (j) sodium acetate
17. Give the correct name for each of the following:
    (a) $Al(OH)_3$ (d) $NaOH$ (g) $(NH_4)_3PO_4$
    (b) Li_2SO_4 (e) NH_4I (h) $BeCO_3$
    (c) $Sr(HCO_3)_2$ (f) $Mg(NO_3)_2$ (i) $BaSO_4$
                                      (j) $KHSO_4$

11 RECIPE FOR A SOLUTION

Many chemists regularly work with solutions.

CONTENTS

11.1 Finding a Solution

Solutions, as you remember, are mixtures that look as if they are pure substances. Air, for example, is a solution. Approximately 80% of dry air is nitrogen gas, making nitrogen the solvent or majority component. Oxygen, which comprises about 19% of dry air, is the most important solute or minority component. The gas carbon dioxide is an example of other solutes that are present in much smaller quantities.

One characteristic of a solution is variable composition. The quality of air varies from place to place because the proportions of its components vary. You are very familiar with one variation, namely, the proportion of water vapour in the air. On very humid days, the proportion of water vapour in the air is higher than on very dry days. Another variable that affects air quality is the quantity of various pollutants. These may be irritating and sometimes smelly.

Steel is an example of a solid solution called an alloy. In "carbon steel," iron is the solvent and carbon is the solute. By varying the proportion of carbon and modifying the production method, industry produces a wide range of steels with different degrees of strength and flexibility. Unfortunately, carbon steel rusts, or corrodes, at much the same rate as pure iron does.

The rate of rusting may be reduced by 80% with the addition of only 5% chromium. An alloy containing 16% chromium is a virtually rust-free form of steel, called "stainless steel." However, the higher percentage of chromium also makes the steel a poorer conductor of heat. One drawback of stainless steel cookware is that "hot spots" or regions of uneven heat distribution may develop. What advantages does stainless steel have over cast iron for cookware?

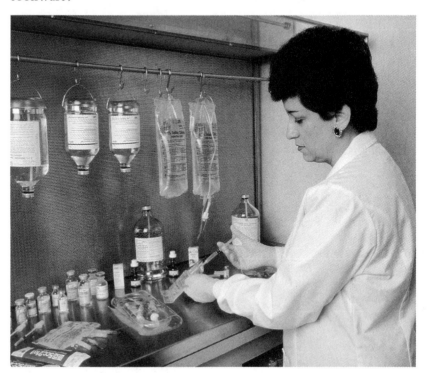

Fig. 1 *This pharmacist is preparing bags of saline transfusions.*

Fig. 2 *Solders made from different proportions of tin, lead, and bismuth have different melting points.*

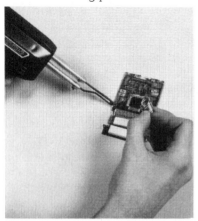

We usually think of solutions as liquids. The salt or saline solution used in hospitals for transfusing drugs into the blood stream is called "normal" if it is 0.9% sodium chloride (Figure 1). Body fluids, like blood, also contain 0.9% sodium chloride. If a saline transfusion were too salty or not salty enough, a patient could be seriously harmed.

Questions

1. Check around your home and find as many solutions as possible. You can find solutions in the kitchen, in the medicine cabinet, in food, almost anywhere in your home! For each solution — solid, liquid, or gas — try to list its solvent and solute(s).

2. (a) Where did you find the name of each commercial solution, that is, each solution bought from a store?
   (b) Were you able to find the proportion of each solvent and solute in each solution? Explain.

Fig. 3 *What does windshield antifreeze contain?*

11.2 Mass Percent

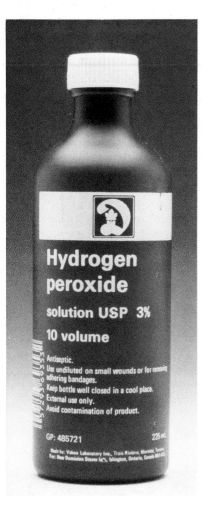

Up to this point, we have mainly thought about what is in a solution, rather than about how much of the solute there is in a solution. The proportion of a solute is important information and is called the **concentration** of a solution. We may not use the word "concentration" in ordinary conversation, but we often encounter this concept in everyday life. The next time you ask for strong black coffee or hear a news report about humidity on a steamy summer day, keep in mind that you are really thinking about the concentration of a solution. There are many ways of expressing concentration. How many different ways of expressing concentration did you encounter when answering the questions in section 11.1?

Look at the picture of the bottle of hydrogen peroxide (Figure 4). The label says that the solution contains 3% hydrogen peroxide. What do you think "3%" means?

One of the simplest recipes for a solution tells us how many grams of solute to dissolve in 100 g of solution. This way of expressing concentration is called the **mass percent**.

$$\text{Mass percent} = \frac{\text{mass of solute}}{\text{mass of solution}} \times 100\%$$

The solution labelled "3% hydrogen peroxide" contains 3 g of hydrogen peroxide in every 100 g of the solution. What makes up the other 97% of the solution?

Fig. 4 *Solutions containing 3% to 6% hydrogen peroxide are used as antiseptics. A strong hydrogen peroxide solution may damage your skin. If it gets on your skin, wash it off with water.*

Questions

1. Solution A contains 10 g of table salt in 90 g of water. Solution B contains 20 g of table salt in 80 g of water.
   (a) Which solution is more concentrated?
   (b) What is the concentration of each solution, expressed as a mass percent?

2. If you look at the label on a container of vinegar, you will see that the vinegar contains 5% acetic acid *by volume*. You already know that 5% by mass means that there is 5 g of solute in every 100 g of solution. What do you think 5% by volume means?

| Investigation | ## 11.3 Preparing a Solution of Known Mass Percent |

In section 11.2, you met the equation for mass percent:

$$\text{Mass percent} = \frac{\text{mass of solute}}{\text{mass of solution}} \times 100\%$$

You can now apply this equation in a practical setting. Before you begin this investigation, study the following example.

Example 1

In a 0.90% sodium chloride solution, there are 0.90 g of sodium chloride and 99.10 g of water for every 100.00 g of solution.
(a) What mass of salt would you need to prepare 500.00 g of the "normal" saline solution?
(b) What mass of water would you need?

Solution

(a)
$$\text{Mass percent} = \frac{\text{mass of solute}}{\text{mass of solution}} \times 100\%$$

$$\text{So,} \quad 0.90\% = \frac{\text{mass of solute}}{500.00 \text{ g}} \times 100\%$$

$$\frac{0.90\% \times 500.00 \text{ g}}{100\%} = \text{mass of solute}$$

$$4.5 \text{ g} = \text{mass of solute}$$

You would need 4.5 g of salt.
(b) Mass of water = 500.00 g − 4.5 g
           = 495.5 g

In this investigation, you will be preparing a solution of a specific concentration. Your teacher will indicate:

• the coloured solute you are to use
• the target concentration
• the mass of the solution

The solvent you will use is water, which has a convenient property: its density is 1.0 g/mL. Rather than measuring a certain number of grams of water, you may choose, instead, to measure an equal number of millilitres.

Materials

List the equipment you will need for weighing the solute, transferring it to a flask, and adding a measured mass or volume of water. Ask your teacher to check and approve your list.

Method

1. List the steps you must take to prepare the solution. Ask your teacher to check and approve your list.

2. Prepare the solution.

3. Label the solution with the name of the solute and the solution concentration. Express the concentration as a mass percent.

Follow-up

1. Compare your solution with other solutions made by your classmates.
   (a) Which solutions look most concentrated?
   (b) How can you tell?
   (c) Does your judgment of relative concentrations agree with the labels on the containers?

2. Suppose you wish to prepare 250 g of a solution that has a concentration of 4.0% by mass.
   (a) What masses of solute and solvent do you need?
   (b) If the solvent is water, what volume of water will you use?

| Investigation | ## 11.4 Concentration in Grams per Litre |

Because it is easier and faster to measure the volume of a liquid than to measure its mass, there is another way of expressing the concentration of a solution. The concentration may be expressed as grams of solute per litre of solution.

$$\text{Concentration (g/L)} = \frac{\text{mass of solute (g)}}{\text{volume of solution (L)}}$$

In a 3 g/L solution of calcium chloride, there are 3 g of calcium chloride and enough water to make 1 L of solution. To prepare such a solution, we use a special kind of flask. A volumetric flask is designed to contain exactly the required volume of solution and is shaped to make mixing easier (Figure 5). Treat volumetric flasks with care; they are fragile and expensive.

Fig. 5 *A volumetric flask.*

A **meniscus** is the curved surface of a column of liquid.

Fig. 6 *When your eye is level with the circular mark, the mark and the bottom of the meniscus should coincide.*

To convert millilitres to litres, divide by 1000.

Don't forget your safety goggles!

Volumetric flasks come in different sizes (100 mL, 250 mL, and so on). If you look at a volumetric flask, you will see that it has a long neck with a circular mark part way up it. When you have filled the flask, make sure the bottom of the meniscus is exactly level with the circular mark (Figure 6).

Your teacher will show you how to use a volumetric flask correctly.

Example 1

What mass of sugar do you need to make 300 mL of a 5.00 g/L sugar solution?

Solution

$$\text{Concentration (g/L)} = \frac{\text{mass of solute (g)}}{\text{volume of solution (L)}}$$

$$5.00\,\text{g/L} = \frac{\text{mass of solute}}{0.300\,\text{L}}$$

$$0.300\,\text{L} \times \frac{5.00\,\text{g}}{1\,\text{L}} = \text{mass of solute}$$

$$1.50\,\text{g} = \text{mass of solute}$$

You will need 1.50 g of sugar.

Now, ask your teacher for the volume and concentration of the target solution and calculate the mass of coloured solute you will need.

Materials

List the materials and equipment that you will need. Ask your teacher to check and approve your list.

Method

1. List the steps you must take to prepare your solution.

2. Show the list to your teacher for approval, and then go ahead.

3. Do not forget to label your solution with the name of the solute and the concentration in grams per litre. Keep the solution for section 11.5 and investigation 11.6.

Questions

1. Why are mass percent and grams per litre convenient concentration units for manufacturing and industrial uses?

2. Arrange each of the following sets of solutions in order of *increasing* concentration (that is, with the lowest concentration first):
   (a) 0.2% NaCl; 1.25% NaCl; 0.04% NaCl
   (b) 4 g/L HCl; 0.33 g/L HCl; 10 g/L HCl
   (c) 3% NaOH; 0.3% NaOH; 5 g NaOH/100 g solution;
      5 g NaOH + 45 g H_2O
   (d) 0.500 g/L; 5 mg/L; 0.250 g solute in 250 mL solution;
      0.100 g solute in 500 mL solution

3. What mass of solute do you need to make 50.0 mL of a 20.0 g/L solution?

4. Is it possible to prepare 1 L of a 100 g/L sodium chloride solution by dissolving 100 g of sodium chloride in 1 L of water? Explain your answer.

11.5 Diluting a Solution

Sometimes the solution on hand is too concentrated for its intended use. So you have to "water down" or **dilute** the solution. When you are preparing orange juice from frozen concentrate, you dilute the contents of the can with three or more cans of water. When you add hot water to a few squirts of liquid dish detergent, you are diluting the detergent. What are other examples of dilution that you encounter in your daily life?

Pour out 10 mL of the coloured solution from investigation 11.4 into a beaker. Add another 10 mL of water to this sample. What is the concentration of this new solution? How did you calculate it?

Now add another 10 mL of water to the diluted solution. Now, what is the concentration? How did you calculate it?

Example 1

You need 100 mL of 1.0 g/L HCl (hydrochloric acid), and all you have in stock is a bottle labelled 5.0 g/L HCl. How can you prepare the solution you need?

Solution

The 5.0 g/L solution is five times more concentrated than you need, so you must add water to dilute a sample of it. The key to solving this problem is to realize that, in adding water to a solution, you will not change the quantity of solute present. The first step, then, is to calculate the mass of solute in 100 mL of 1.0 g/L HCl.

$$\text{Concentration (g/L)} = \frac{\text{mass of solute (g)}}{\text{volume of solution (L)}}$$

$$1.0 \ \text{g/L} = \frac{\text{mass of solute}}{0.100 \, \text{L}}$$

$$0.100 \, \text{L} \times \frac{1.0 \, \text{g}}{1 \, \text{L}} = \text{mass of solute}$$

$$0.10 \, \text{g} = \text{mass of solute}$$

As the mass of solute is unchanged by dilution, the sample of 5.0 g/L solution, before dilution, must also contain 0.10 g of solute. We can now calculate the volume of that sample.

$$\text{Concentration (g/L)} = \frac{\text{mass of solute (g)}}{\text{volume of solution (L)}}$$

$$5.0 \, \text{g/L} = \frac{0.10 \, \text{g}}{\text{volume of solution}}$$

$$\text{volume of solution} = \frac{0.10 \, \text{g}}{5.0 \, \text{g/L}}$$

$$= 0.020 \, \text{L}$$

$$= 20 \, \text{mL}$$

The units may be easier to understand if you think of $\dfrac{0.10 \, \text{g}}{5.0 \, \text{g/L}}$ as $0.10 \, \text{g} \times \dfrac{1 \, \text{L}}{5.0 \, \text{g}}$.

To convert litres to millilitres, multiply by 1000.

Thus, you must transfer 20 mL of the 5.0 g/L solution into a volumetric flask and then add enough water so that you will have exactly 100 mL of 0.10 g/L solution.

Alternative Solution

The basis of the above calculation is the fact that the mass of solute in a solution does not change when water is added. But, mass of solute (g) = concentration (g/L) × volume of solution (L). Therefore, it must be true that the value of concentration × volume must be the same for both the concentrated and dilute solutions. This relationship is sometimes written as:

$$(C \times V)\, \text{concentrated} = (C \times V)\, \text{dilute}$$

$$\text{Thus,} \quad 5.0\,\text{g/L} \times V\, \text{concentrated} = 1.0\,\text{g/L} \times 0.100\,\text{L}$$

$$V\, \text{concentrated} = \frac{1.0\,\text{g/L} \times 0.100\,\text{L}}{5.0\,\text{g/L}}$$

$$= 0.020\,\text{L}$$

$$= 20\,\text{mL}$$

Questions

1. What volume of a 9.0 g/L solution do you need to prepare 1.2 L of a 1.5 g/L solution?

2. What volume of a 2.4 g/L solution can you prepare from 200 mL of a 6.0 g/L solution?

Investigation | 11.6 *Preparing a Dilute Solution*

Fig. 7 *The correct use of a pipette to transfer an exact volume of liquid to another container.*

Using the techniques shown in Figures 7 and 8, you will prepare 100 mL of a target solution from the solution of known concentration you prepared in investigation 11.4.

(a) Partly fill the pipette with distilled water.

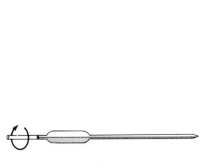

(b) Turn the pipette to rinse the inside thoroughly with the distilled water.

(c) Discard the distilled water.

311

Fig. 7 *(continued)*

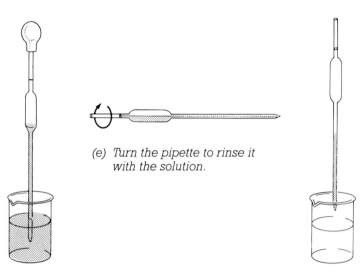

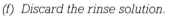

(e) Turn the pipette to rinse it with the solution.

(g) Fill the pipette above the line with the solution.

(d) Partly fill the pipette with the solution to be used.

(f) Discard the rinse solution.

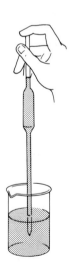

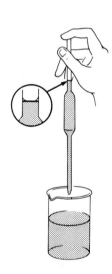

(j) Transfer the pipette to an empty vessel. Place the tip of the pipette along the inside of the vessel. Remove your forefinger from the top of the pipette and allow the liquid to drain out. Wait about 10 s before removing the pipette and always leave the last drop in the pipette.

(h) Remove the bulb and quickly replace it with a moist forefinger.

(i) Lift the pipette out of the liquid. Slowly raise the inside edge of the finger and allow the liquid to drain until the bottom of the meniscus just touches the line.

Fig. 8 *Diluting a solution.*

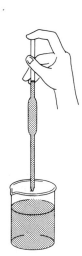

(a) Fill a pipette with the
    concentrated solution
    given.

(b) Use the pipette to
    transfer a known
    volume of the solution
    into a volumetric flask.

(c) Swirl the contents as
    distilled water is added
    to the flask until near
    the mark.

(d) Mix well by capping
    the flask and shaking.

(e) Then add distilled
    water to the mark.

(f) Again mix well.

Don't forget your safety goggles!

Materials

After studying the techniques of dilution, list all the equipment you will need and ask your teacher to check and approve your list.

Method

1. Calculate the volume of concentrated solution you will need to prepare 100 mL of the dilute solution.

2. List the steps you must take to dilute your solution properly.

3. After getting your teacher's approval, carry out the dilution.

Follow-up

1. How does the appearance of your dilute solution compare to that of the concentrated one?

A **standard solution** is a solution of accurately known concentration.

2. Your teacher will show you a standard solution of the same concentration as your dilute solution should have. Compare the colour of your solution to the standard. Can you tell by only visual inspection how accurate you were in your dilution?

3. How many grams of solute are there in each of the following quantities of solution?
   (a) 100 g of 4.0% sugar solution
   (b) 300 g of 0.50% sodium chloride solution
   (c) 1500 g of 3.00% silver nitrate solution
   (d) 2.00 kg of 5.00% acetic acid solution
   (e) 100 mL of 5.00 g/L acetic acid solution
   (f) 250 mL of 3.00 g/L silver nitrate solution
   (g) 1500 mL of 4.0 g/L sugar solution
   (h) 2.00 L of 0.50 g/L sodium chloride solution

4. Explain how you would prepare 500 mL of 0.50 g/L sugar solution from each of the following solutions.
   (a) 5.0 g/L sugar solution
   (b) 2.5 g/L sugar solution
   (c) 0.75 g/L sugar solution

11.7 Measuring Quantity

How do you measure the quantity of hamburger meat you will need to feed all of your friends for lunch? the quantity of air in an automobile tire? the quantity of paint to cover the walls of your room? The unit of measurement depends very much on the physical state — solid, liquid, or gas.

How many different ways of measuring these quantities can you think of? The quantity of hamburger meat could be expressed by means of its mass in kilograms (kg). We can describe the quantity of air in a tire by means of its pressure in kilopascals (kPa). We may measure the quantity of paint by means of its volume in litres (L). In what units did you record the ways of measuring that you thought of? What are the usual symbols for those units?

Suppose you wanted to find out how many molecules of nitrogen (N_2) are in the air trapped in the tire. How would you measure this quantity? How would you measure the number of oxygen atoms that combine with 1 g of magnesium atoms to form magnesium oxide (MgO)? Discuss these questions in small groups. Share your proposals with your classmates.

One difficulty you may have anticipated results from the tiny sizes of atoms and molecules. Whatever methods you proposed for measuring the numbers of nitrogen molecules and oxygen atoms, you should be aware of this difficulty, namely, that the numbers are enormous. So, chemists are faced with the need for a unit that allows them to work with enormous numbers.

The unit which chemists use to express the number of particles in a sample of matter is called the mole. A mole represents a fixed number of particles in the same way that a pair always represents two items and a dozen always represents twelve items. Can you think of other words that stand for fixed numbers of items?

The symbol for mole is "mol."

Your teacher will now show you a series of beakers, each containing a solid or a liquid. Each beaker contains 1 mol of a substance. It may surprise you to see that 1 mol of one substance does not necessarily have the same volume as 1 mol of another substance. Your teacher will tell you the mass of the substance in each beaker.

Gases have an interesting property that makes them different from solids and liquids. One mole of a specific gas occupies about the same volume as 1 mol of another gas, provided that both gases are at the same temperature and under the same pressure. For any gas at room temperature and atmospheric pressure, the volume of 1 mol is about 25 L, for example. Your teacher will show you what a volume of 25 L looks like.

Questions

1. Was there any pattern to the masses or volumes of 1-mol quantities of the various solids and liquids?

2. One mole of a solid or liquid does not usually have the same mass or volume as 1 mol of another solid or liquid. However, 1 mol of a gas has about the same volume as 1 mol of another gas, under the same conditions of temperature and pressure. Can you explain this?

3. You have not, so far, compared the masses of 1-mol quantities of various gases.
   (a) Can you propose a method for measuring these masses?
   (b) Can you suggest any difficulties that might arise in your proposed method?

11.8 The Mole

In section 11.7, the concept of the mole was introduced. Chemists use the mole in discussing the number of particles in a sample of matter. The volume and mass of 1 mol of a substance vary from one substance to another.

The mole is a huge number, 602 000 000 000 000 000 000 000, which is more convenient to write in scientific or exponential notation as 6.02×10^{23}. This number is usually called Avogadro's number, after the Italian chemist Amadeo Avogadro.

In order to identify the type of particle present in a substance, we have only to look at its formula. For example,

A **mole** means 6.02×10^{23}, as a dozen means 12.

- the formulas of elements, such as Mg, Al, Si, Ne, and O, represent *atoms*;
- other formulas of elements, such as H_2, O_2, P_4, and S_8, represent *molecules*;
- the formulas of compounds, such as H_2O, $CaCO_3$, NH_3, and NaCl, also represent molecules.

We can use chemical formulas to interpret the meaning of a mole for various substances:

$$1.00 \text{ mol of magnesium (Mg)} = 6.02 \times 10^{23} \text{ atoms of magnesium}$$
$$1.00 \text{ mol of aluminum (Al)} = 6.02 \times 10^{23} \text{ atoms of aluminum}$$
$$1.00 \text{ mol of oxygen (O)} = 6.02 \times 10^{23} \text{ atoms of oxygen}$$
$$1.00 \text{ mol of oxygen (O}_2) = 6.02 \times 10^{23} \text{ molecules of oxygen}$$
$$1.00 \text{ mol of hydrogen (H}_2) = 6.02 \times 10^{23} \text{ molecules of hydrogen}$$
$$1.00 \text{ mol of phosphorus (P}_4) = 6.02 \times 10^{23} \text{ molecules of phosphorus}$$
$$1.00 \text{ mol of water (H}_2O) = 6.02 \times 10^{23} \text{ molecules of water}$$
$$1.00 \text{ mol of ammonia (NH}_3) = 6.02 \times 10^{23} \text{ molecules of ammonia}$$
$$1.00 \text{ mol of sodium chloride (NaCl)} = 6.02 \times 10^{23} \text{ molecules of sodium chloride}$$

AMADEO AVOGADRO

At the time Dalton published his atomic theory, the French chemist, J. L. Gay-Lussac, and others were studying the relationship between volumes of gases involved in chemical reactions. They observed that, when hydrogen gas combines with oxygen gas to form water, two volumes of hydrogen combined with one volume of oxygen to give exactly two volumes of gaseous water. Similar whole number ratios were found in other reactions between gases.

In 1811, through a brilliant stroke of intuition, the Italian chemist Lorenzi Romeo Amadeo Carlo Avogadro, Count of Quarengna and Correto, discovered the simple key to understanding Gay-Lussac's observations. He showed that the fixed volume ratios could only occur if each volume of gas contained an equal number of gaseous atoms or molecules. If, as Dalton pointed out, atoms combined in whole number ratios to form molecules, then volumes containing equal numbers of atoms would also combine in whole number ratios. Avogadro's hypothesis was opposed even by such influential chemists as Dalton himself, but the idea was eventually accepted and a great deal of chemistry has developed from it.

Fig. 9 *Avogadro considered the numbers of atoms or molecules in volumes of gases.*

Questions

1. How many particles are present in one mole of each of the following? In each case, specify whether the particles are atoms or molecules.
   (a) sulfur (S_8)
   (b) uranium (U)
   (c) neon (Ne)
   (d) lithium bromide (LiBr)

2. (a) Which is heavier, one mole of oxygen atoms or one mole of oxygen molecules?
   (b) By what factor do the masses of one mole of oxygen atoms and one mole of oxygen molecules differ? Explain.

3. Suppose that the entire population of Canada (about 27 000 000 people) were hired to count 1 mol of objects. We share out the objects and all work steadily, counting one object per second for 40 h each week. At this rate, assuming no vacations, how long would it take us to count all the objects? (Answer in years.)

The symbol for hour is h.

11.9 Moles Within Moles

Fig. 10 *Can elephants teach us about moles?*

If you see a dozen elephants, how many dozen elephant tusks can you count? how many dozen elephant trunks? how many dozen elephant legs? how many dozen elephant tails? Your answers depend on a knowledge of an elephant's anatomy. You must know how many trunks, tusks, legs, and tails one elephant has. How are these questions about elephants related to chemistry?

We can similarly consider the number of atoms of each element in a certain number of molecules, but only if we know a single molecule's "anatomy." This is specified in the chemical formula. For example, the chemical formula for an oxygen molecule is O_2. How many oxygen atoms are there in one oxygen molecule? in two oxygen molecules? in a dozen oxygen molecules? As you can see in each case, we simply multiply the number of oxygen molecules by 2 to get the number of oxygen atoms.

The previous paragraph illustrates a familiar interpretation of a chemical formula. The subscript(s) in a formula tell us the number of atoms of each element in a molecule. But there is another possible interpretation. Suppose you were given 6.02×10^{23} oxygen molecules (O_2), then how many oxygen atoms are there altogether? The answer is clearly $2 \times 6.02 \times 10^{23}$. Now, let's translate this information into moles:

$$6.02 \times 10^{23} \text{ oxygen molecules} = 1 \text{ mol of oxygen molecules}$$
$$2 \times 6.02 \times 10^{23} \text{ oxygen atoms} = 2 \text{ mol of oxygen atoms}$$

In other words, the formula O_2 indicates that 1 mol of oxygen molecules contains 2 mol of oxygen atoms. In general, the subscripts in a chemical formula show the number of moles of atoms of each element in 1 mol of molecules.

We can apply this interpretation to all kinds of molecules. For example, in 1 mol of water (H_2O), there are 2 mol of hydrogen atoms (H) and 1 mol of oxygen atoms (O). How many moles of oxygen atoms are in 1 mol of calcium carbonate molecules ($CaCO_3$)? How many moles of hydrogen atoms are in 1 mol of ammonia molecules (NH_3)? How many individual hydrogen atoms are in 1 mol of ammonia molecules?

When we deal with more or less than 1 mol of molecules, we can still calculate the numbers of moles of atoms present. For example, in 0.5 mol of water (H_2O), there are 1 mol of hydrogen atoms (H) and 0.5 mol of oxygen atoms (O). In 4 mol of water, how many moles of hydrogen atoms (H) are there? how many moles of oxygen atoms (O)? how many individual oxygen atoms (O)?

Questions

1. Copy and complete the following chart in your notebook. (Please do not write in your textbook.)

Substance	Quantity	Number of moles of O atoms	Number of O atoms
ozone, O_3	1.00 mol		
carbon dioxide, CO_2	1.00 mol		
sulfur trioxide, SO_3	1.00 mol		
calcium sulfate, $CaSO_4$	1.00 mol		
aluminum oxide, Al_2O_3	1.00 mol		
barium nitrate, $Ba(NO_3)_2$	1.00 mol		
aluminum nitrate, $Al(NO_3)_3$	1.00 mol		

2. Copy and complete the following chart in your notebook. (Please do not write in your textbook.)

Substance	Quantity	Number of moles of N atoms	Number of N atoms
N_2O_4	3.00 mol		
N_2O_4	0.500 mol		
N_2O_4	0.750 mol		
KNO_3	2.00 mol		
$Al(NO_3)_3$	1.50 mol		
NH_4NO_3	2.50 mol		

11.10 Molar Mass

Your only awareness of the mole, so far, is in terms of a number of objects. If you completed section 11.8, you know that the number of objects in 1 mol is 6.02×10^{23}, which we call Avogadro's number. Now suppose that you were told to purchase 1 mol of table salt (sodium chloride) at the supermarket. How would you do this? Why is it not possible just to count Avogadro's number of sodium chloride molecules and put them into a container? List all the reasons you can think of and discuss them with your group.

6.02×10^{23} grains of rice would cover the province of Québec to a height of nearly 8 km.

If it is not possible to count the number of particles you need, what other method(s) might you use for obtaining 1 mol of sodium chloride? Is there any additional information you need about how the mole is defined? Do you think that your proposed method(s) would work equally well for all states of matter? Discuss your answers with your group.

Consider how you do, in fact, buy table salt at the supermarket. What units are printed on the box that indicate the quantity of salt inside? The units on the box suggest that there is an easier way to measure a mole of a substance than by counting out the individual atoms or molecules. One mole of atoms of an element in fact has a mass in grams that is numerically equal to the atomic mass of that element. The mass of 1 mol of atoms of an element is sometimes called the "gram atomic mass" of the element, but the preferred term is **molar mass**. Molar mass is expressed as grams per mole, or g/mol.

As you may remember from sections 4.1 and 4.6, the atomic mass is the average mass of the naturally occurring isotopes of an element. Where would you look up the atomic mass of an element?

Thus, the molar mass of carbon atoms is 12.0 g/mol. The molar mass of potassium atoms is 39.1 g/mol. What is the molar mass of magnesium atoms? iodine atoms? oxygen atoms?

To calculate the molar mass of a compound, add up the molar masses of the elements shown in the molecular formula. As the molar masses of the individual elements are all expressed in g/mol, the molar mass of a compound must also be in g/mol.

Example 1

Calculate the molar mass of calcium carbonate, $CaCO_3$.

Solution

The formula indicates the presence of 1 mol of calcium atoms, Ca, 1 mol of carbon atoms, C, and 3 mol of oxygen atoms, O, in 1 mol of calcium carbonate, $CaCO_3$.

To determine the mass of 1 mol of $CaCO_3$, add the masses of 1 mol Ca, 1 mol C, and 3 mol O.

$$
\begin{aligned}
1\,\text{mol Ca} &= 40.1\,\text{g} \\
1\,\text{mol C} &= 12.0\,\text{g} \\
3\,\text{mol O} &= 3 \times 16.0\,\text{g} \\
&= 48.0\,\text{g} \\
1\,\text{mol CaCO}_3 &= 40.1\,\text{g} + 12.0\,\text{g} + 48.0\,\text{g} \\
&= 100.1\,\text{g}
\end{aligned}
$$

So, the molar mass of calcium carbonate is 100.1 g/mol.

Example 2

What is the mass, in grams, of two moles of calcium carbonate?

Solution

The mass must be twice the mass of one mole, as determined in Example 1. So, the mass of two moles of calcium carbonate is 200.2 g.

If you examine the solution to Example 2, you will see that we calculated a mass by multiplying a molar mass by a number of moles. We can write an equation to reflect this operation:

mass of substance (g) =
number of moles of substance (mol) × molar mass of substance (g/mol)

The number of moles is usually given the symbol n, so we can write:

$$n = \frac{\text{mass of substance}}{\text{molar mass of substance}}$$

Example 3

How many moles of fluorine atoms are present in 3.80 g of fluorine atoms?

Solution

From the periodic table, the molar mass of fluorine atoms is 19.0 g/mol.

$$n = \frac{\text{mass of substance (g)}}{\text{molar mass of substance (g/mol)}}$$

$$= \frac{3.80\,\text{g}}{19.0\,\text{g/mol}}$$

$$= 0.200\,\text{mol}$$

There are 0.200 mol of fluorine atoms present.

The units may be clearer if you think of this as $3.80\,\text{g} \times \dfrac{1\,\text{mol}}{19.0\,\text{g}}$

Example 4

How many moles of water are in 100 g of water?

Solution

The molecular formula of water is H_2O.
The molar mass $= 2(1.0\,g/mol) + 16.0\,g/mol$

$$= 18.0\,g/mol$$

$$n = \frac{100\,g}{18.0\,g/mol}$$

$$= 5.55\,mol$$

So, there are 5.55 mol of water.

Questions

1. What is the molar mass of atoms of each of the following elements?
   (a) sulfur   (f) cobalt
   (b) silicon   (g) silver
   (c) boron   (h) bromine
   (d) zinc   (i) phosphorus
   (e) lead   (j) manganese

2. Calculate the mass of a mole of each of the following elements or compounds.
   (a) ammonia, NH_3
   (b) oxygen gas, O_2
   (c) carbon dioxide, CO_2
   (d) dinitrogen pentoxide, N_2O_5
   (e) sodium hydrogen carbonate, $NaHCO_3$
   (f) magnesium sulfate, $MgSO_4$
   (g) zinc nitrate, $Zn(NO_3)_2$
   (h) ammonium sulfate, $(NH_4)_2SO_4$

3. Calculate the mass of the following.
   (a) 0.500 mol of ammonia
   (b) 0.750 mol of oxygen gas
   (c) 2.50 mol of carbon dioxide
   (d) 0.100 mol of dinitrogen pentoxide
   (e) 1.25 mol of sodium hydrogen carbonate
   (f) 0.0500 mol of magnesium sulfate
   (g) 0.333 mol of zinc nitrate
   (h) 1.10 mol of ammonium sulfate

4. Copy and complete the following chart in your notebook. (Please do not write in your textbook.)

Substance	Formula	Molar mass of substance (g/mol)	Mass (g)	Number of moles of atoms or molecules
silicon			2.80	
carbon monoxide				0.300
calcium carbonate			25.0	
zinc				2.00
sodium hydroxide			8.00	
aluminum oxide				0.300
magnesium chloride			0.953	

5. The mass of a single molecule is known as its **molecular mass**. Knowing the formula of a molecule, you can determine its molecular mass by adding the atomic masses of all the atoms in the molecule. Since atomic masses are expressed in atomic mass units, symbol u, so are molecular masses.
   (a) Determine the molecular mass of ammonia, NH_3.
   (b) Determine the molar mass of ammonia, NH_3.
   (c) The molar mass of a molecular substance is sometimes referred to as its "gram molecular mass." Explain why this term is used, on the basis of your answers to (a) and (b).

11.11 Molar Concentration of a Solution

In sections 11.2 and 11.4, you learned to express the concentration of a solution in terms of mass percent and grams per litre. How many ways can you think of for expressing solution concentrations using the mole as a unit? Compare your answers to those of your classmates.

Chemists often express the concentration of a solution in moles of solute per litre of solution, symbol mol/L. A concentration expressed in this way is called the **molar concentration**, and is often referred to as the "molarity."

In the SI, the mole is the unit used to express the amount of a substance. The word "amount" should only be used for a quantity expressed in moles.

$$\text{Molar concentration (molarity)} \ = \ \frac{\text{amount of solute (mol)}}{\text{volume of solution (L)}}$$

Example 1

If 8 g of sodium hydroxide is dissolved in 100 mL (0.100 L) of solution, what is the molar concentration of the solution?

Solution

The formula of sodium hydroxide is NaOH.
The molar mass of NaOH is 40.0 g/mol.

$$\text{Amount of NaOH (mol)} \ = \ \frac{\text{mass of NaOH}}{\text{molar mass of NaOH}}$$

$$= \ \frac{8\,g}{40.0\,g/mol}$$

$$= \ 0.2\,mol$$

$$\text{Molar concentration} \ = \ \frac{\text{amount of solute (mol)}}{\text{volume of solution (L)}}$$

$$= \ \frac{0.2\,mol}{0.100\,L}$$

$$= \ 2\,mol/L$$

The molar concentration is 2 mol/L.

Example 2

Suppose you need 5.00 L of a 0.100 mol/L sodium nitrate solution. What mass of sodium nitrate must you use?

Solution

We can first calculate the amount of solute, in mol:

$$\text{Molar concentration} \ = \ \frac{\text{amount of solute (mol)}}{\text{volume of solution (L)}}$$

$$0.100\,mol/L \ = \ \frac{\text{amount of solute}}{5.00\,L}$$

$$\frac{0.100\,mol}{1\,L} \times 5.00\,L \ = \ \text{amount of solute}$$

$$0.500\,mol \ = \ \text{amount of solute}$$

We can now consider the mass of solute needed.
The formula of sodium nitrate is $NaNO_3$.
The molar mass of $NaNO_3$ is 85.0 g/mol.

$$\text{Amount of } NaNO_3 = \frac{\text{mass of } NaNO_3}{\text{molar mass of } NaNO_3}$$

$$0.500 \, \text{mol} = \frac{\text{mass of } NaNO_3}{85.0 \, \text{g/mol}}$$

$$0.500 \, \text{mol} \times 85.0 \, \text{g/mol} = \text{mass of } NaNO_3$$

$$42.5 \, \text{g} = \text{mass of } NaNO_3$$

Therefore, you must use 42.5 g of sodium nitrate.

Example 3

Suppose you want to make as large a volume of 5.00 mol/L sodium chloride solution as possible. You have a bottle that contains 1.17 kg of this substance. What volume of solution can you make?

Solution

The formula of sodium chloride is NaCl.
The molar mass of NaCl is 58.5 g/mol.

To convert kilograms to grams, multiply by 1000.

$$\text{Amount of available NaCl (mol)} = \frac{\text{mass of NaCl}}{\text{molar mass of NaCl}}$$

$$= \frac{1.17 \times 10^3 \, \text{g}}{58.5 \, \text{g/mol}}$$

$$= 20.0 \, \text{mol}$$

$$\text{Molar concentration} = \frac{\text{amount of solute (mol)}}{\text{volume of solution (L)}}$$

$$5.00 \, \text{mol/L} = \frac{20.0 \, \text{mol}}{\text{volume of solution}}$$

$$\text{volume of solution} = \frac{20.0 \, \text{mol}}{5.00 \, \text{mol/L}}$$

$$= 4.00 \, \text{L}$$

So, you can make 4.00 L of solution.

Questions

Remember, 1000 mL = 1 L.

1. Calculate the molar concentration of each of the following solutions:
   (a) 5.85 g of sodium chloride dissolved in 2.0 L of solution
   (b) 5.61 g of potassium hydroxide dissolved in 100 mL of solution
   (c) 0.741 g of calcium hydroxide dissolved in 100.0 mL of solution

2. What mass of sodium chloride must be used in order to make up 100 mL of a 0.20 mol/L solution? Copy and complete the calculations below in your notebook. (Please do not write in your textbook.)

   The formula of sodium chloride is NaCl.

   Molar mass of NaCl = ? g/mol

   $$\text{Molar concentration} = \frac{\text{amount of solute}}{\text{volume of solution}}$$

   So, amount (mol) = molarity (mol/L) × volume (L)

   Amount of NaCl = 0.20 mol/L × 0.10 L

   = ? mol

   $$\text{Amount of NaCl} = \frac{\text{mass of NaCl}}{\text{molar mass of NaCl}}$$

   amount of NaCl × molar mass of NaCl = mass of NaCl

   Therefore, mass of NaCl = ? mol × ? g/mol
   = ? g

3. Calculate the mass of solute in each solution.
   (a) 200 mL of a 0.40 mol/L solution of magnesium nitrate
   (b) 100 mL of a 0.25 mol/L solution of potassium iodide
   (c) 20.0 L of a 0.50 mol/L solution of sodium hydroxide

4. Calculate the maximum volume, in litres, of a 0.100 mol/L solution you can prepare from each of the following:
   (a) 40.0 g of sodium hydroxide
   (b) 73.0 g of hydrogen chloride
   (c) 1.483 g of magnesium nitrate

Recall that a solution of hydrogen chloride in water is called hydrochloric acid.

5. Can you prepare 1.00 L of 2.50 mol/L sodium chloride solution by adding 1.00 L of water to 2.50 mol of sodium chloride? Explain your answer.

<table>
<tr><td>

Investigation

</td><td>

11.12 Preparing a Solution of Known Molar Concentration

</td></tr>
</table>

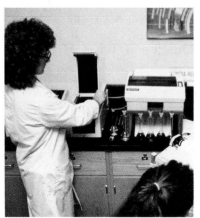

Fig. 11 *Standard solutions are very important in chemistry.*

Don't forget your safety goggles!

Your friend, who is a research chemist, has left a message for you with your chemistry teacher. Your help is needed to prepare a special solution. All that you are given is the formula of a coloured compound, a volumetric flask, and the concentration of the solution that you are to prepare. What do you do first? How much solute do you need?

Materials

List all the equipment you will need for weighing and transferring your solute to the volumetric flask. You will also need a supply of deionized water. Ask your teacher to approve the list before you continue.

Method

1. Write down the steps you must take. This method is similar to the one in investigation 11.4.

2. Ask your teacher to check and approve your method before you begin.

3. Keep your solution for the next investigation.

Follow-up

1. Compare your solution to a standard solution of the required concentration.
   (a) Do the colours of the solutions match?
   (b) If the colours are different, can you tell which solution is more concentrated? Explain your answer.

Investigation

11.13 Diluting a Solution of Known Molar Concentration

The research chemist calls back to tell you that the previous solution you prepared in investigation 11.12 is too concentrated. You are asked to dilute it to one-fifth of the original concentration.

Materials

Don't forget your safety goggles!

List the equipment you will need. Get your teacher's approval before proceeding.

Method

1. Outline how you will prepare the diluted solution. List all the steps you plan to take.

2. Do not begin without your teacher's approval.

Follow-up

1. (a) How does the colour of your diluted solution compare to that of the standard solution of the intended concentration?
   (b) What does this comparison tell you?

2. How many moles of solute were in the volume of concentrated solution that you transferred to the volumetric flask?

3. How many moles of solute were in the volume of dilute solution you prepared?

4. (a) From your answers to questions 2 and 3, is there any relationship between the amount of solute taken out of the concentrated solution and the amount in the dilute solution?
   (b) Write a mathematical equation that could be used in dilutions of solutions. The equation should relate the volume and concentration of the concentrated solution to the volume and concentration of the dilute solution.

Investigation	*11.14 Another Dilute Solution*

There's a frantic knock at the door! It is the research chemist again. This time, you are asked for 100 mL of a solution with a different concentration. This solution is not so easy to prepare. Use the mathematical relationship you developed in investigation 11.13 to plan the preparation of this dilute solution. In any calculation you do, make sure that the answer has the correct units. It is also a good idea to check that the answer is reasonable by making a rough estimate and comparing it to the answer you obtain.

Don't forget your safety goggles!

Materials

Use the same equipment you used in investigation 11.13.

Method

1. List all the steps you must take in the dilution.

2. Ask your teacher to check and approve the list.

Follow-up

1. (a) Does the appearance of the new dilute solution match the standard solution of the intended concentration?
   (b) What conclusion can you draw from your answer to (a)?

2. Calculate the volume of 2.50 mol/L NaOH that you will need to prepare each of the following solutions:
   (a) 250 mL of 1.25 mol/L NaOH
   (b) 1500 mL of 0.750 mol/L NaOH
   (c) 500 mL of 1.00 mol/L NaOH

3. Calculate the molar concentration of the solution prepared by diluting 10.00 mL of 5.00 mol/L HCl to each of the following volumes:
   (a) 250 mL   (c) 1000 mL
   (b) 50.0 mL   (d) 500 mL

4. Calculate the volume of each of these solutions needed to prepare 250 mL of a 0.100 mol/L NaOH solution:
   (a) 1.00 mol/L NaOH   (c) 3.25 mol/L NaOH
   (b) 0.250 mol/L NaOH   (d) 0.850 mol/L NaOH

5. Suppose that you are given 100 mL of a 1.00 mol/L solution. You are asked to prepare 100 mL of a 0.00100 mol/L solution from it. The only glassware available is a 10-mL pipette, a 100-mL volumetric flask, and a few beakers. What procedure would you use?

P O I N T S · T O · R E C A L L

- The proportion of solute in a solution can be expressed as a mass percent:

$$\text{Mass percent (\%)} = \frac{\text{mass of solute (g)}}{\text{mass of solution (g)}} \times 100\%$$

- The mass of solute in a given quantity of solution is also expressed in grams per litre.
- When we add water to a solution in order to lower its concentration, we are diluting the solution. During dilution, the quantity of solute present stays constant.

- To calculate the volume of a concentrated solution needed to prepare a specific volume of dilute solution, use the following equation:

$$(C \times V) \text{ concentrated} = (C \times V) \text{ dilute}$$

- The mole is the unit that represents the amount of a substance.
- A mole of objects contains Avogadro's number (6.02×10^{23}) of objects.
- The molar mass of a substance is found by adding the atomic masses of all the atoms shown

in the chemical formula of the substance and expressing this sum in grams per mole.

• The number of moles, n, is calculated by using:

$$n = \frac{\text{mass of substance (g)}}{\text{molar mass of substance (g/mol)}}$$

• The molar concentration of a solution is expressed in moles per litre:

$$\text{molar concentration (mol/L)} = \frac{\text{amount of solute (mol)}}{\text{volume of solution (L)}}$$

R E V I E W • Q U E S T I O N S

1. Which solution contains more solute: 100 g of 5% (by mass) sugar solution or 100 mL of 5 g/L sugar solution? Show your reasoning.

2. How many grams of solute are needed to prepare the following solutions?
   (a) 150 g of 10% HCl solution by mass
   (b) 250 mL of 10 g/L HCl solution
   (c) 250 mL of 10 mol/L HCl solution

3. Calculate the concentration, in grams per litre, of the diluted solutions prepared by adding:
   (a) 2000 mL of water to 1.00 L of 6.0 g/L sugar solution
   (b) 150 mL of water to 100 mL of 1.6 g/L NaCl solution
   (c) 300 mL of water to 150 mL of 4.0 g/L KI solution

4. (a) What volume of water must be added to 100 mL of a 5.00 g/L HNO_3 solution to dilute it to each of the following concentrations?
   (i) 2.50 g/L
   (ii) 4.00 g/L
   (iii) 0.500 g/L
   (b) What assumption did you make in your answers to (a)?

5. What volume of 10 mol/L HCl is needed to prepare 5.0 L of 2.5 mol/L HCl?

6. Calculate the molar mass of the following:
   (a) sodium phosphate
   (b) aluminum sulfate
   (c) diphosphorus pentasulfide
   (d) oxygen gas (O_2)
   (e) hydrogen sulfate

7. Calculate the number of moles of compound in the following:
   (a) 11.7 g of sodium chloride
   (b) 24.5 g of calcium hydroxide
   (c) 78.7 g of hydrogen nitrate

8. Calculate the concentration (mol/L) for each of the following:
   (a) 4.0 g of sodium hydroxide dissolved in 250 mL of solution.
   (b) 13.65 g of zinc chloride dissolved in 200 mL of solution.

9. In point form, indicate how you would use a 25-mL pipette and a 250-mL volumetric flask to make up a 0.001 mol/L hydrochloric acid solution from a 0.1 mol/L hydrochloric acid solution.

10. How many millilitres of 0.400 mol/L ammonium hydroxide solution are need to prepare 500 mL of 0.100 mol/L solution?

11. What concentration would result from transferring 12.0 mL of 2.50 mol/L sulfuric acid to a 500-mL volumetric flask and diluting to the mark?

12 LEARNING MORE FROM ACIDS AND BASES

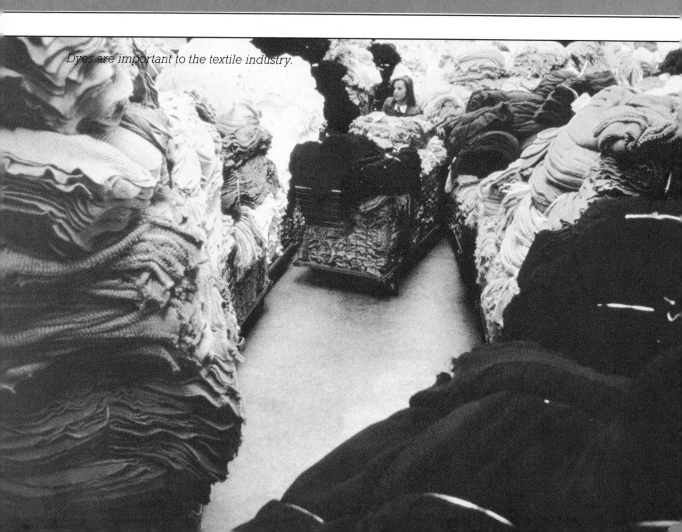

Dyes are important to the textile industry.

CONTENTS

| Investigation | 12.1 Discovering New Indicators |

A chemistry teacher was dismayed when a bright orange spot appeared on her favourite blue wool dress. The spot was caused by a drop of hydrochloric acid. Confident that all she had learned about neutralization was true, she daubed some dilute ammonium hydroxide on the stain. As you might expect, the blue colour returned and her dress was saved. In fact, dyes used in inks and in dyeing fabrics behave as acid–base indicators. They resemble familiar acid–base indicators, such as litmus and phenolphthalein, in the ways they react to acids and bases.

A chemical manufacturing company is quite far behind in its testing. When word reaches the company that your class is interested in studying indicators, the chief of its testing laboratory asks if you might determine the behaviour of some of the company's dyes in acidic and in basic solutions. The company's hope is to market the dyes as acid–base indicators. The results of your testing will assist it in marketing its products.

Compare the new dyes with indicators already on the market — litmus and phenolphthalein. If new indicators are to compete with litmus and phenolphthalein, how well must the new indicators perform? Let's see if they live up to your expectations!

Never use more than a few drops of a chemical indicator to test an acid or base. As an indicator is itself an acid or base, too much of it will affect your test results.

333

Don't forget your safety
goggles!

Bromothymol blue is some-
times called bromthymol blue.

Handle the acid and base with
care.

Many dyes are toxic. Handle
them with care.

Materials

The dyes you will be testing are called methyl orange, methyl red,
and bromothymol blue. For comparison, you will perform the same
tests on the familiar indicators litmus and phenolphthalein. You
will, of course, need a typical acid and a typical base to carry out
your testing. Make a "shopping list" of solutions and equipment
you will need and have the list approved by your teacher.

Method

1. Decide how you will conduct the tests. Test only one dye
   solution at a time. Make it a practice to clean your equipment
   between tests. Then you will not have to worry about mixed-up
   results.

2. Write down a set of directions for yourself and have this
   approved by your teacher.

3. Carry out the procedure and enter all your results in a data
   chart.

Fig. 1 *Fabrics being dyed.*

Follow-up

1. (a) Did the new dyes perform as you had expected?
   (b) In what ways were they similar to litmus and
       phenolphthalein?
   (c) In what ways were they different?

2. (a) Is each of the new dyes capable of distinguishing acids from
       bases?
   (b) Do you regard each of the new dyes as an acceptable acid–
       base indicator? Explain your reasoning.

Investigation

12.2 Introducing the pH Scale

Fig. 2 *To taste good, dairy products must have the right acidity.*

A **buffer** is a solution that prevents large changes in pH when a small amount of acid or base is added to it.

Don't forget your safety goggles!

Handle all acids and bases with care. Some are corrosive.

Almost daily you hear expressions such as "pH-balanced shampoo," the "pH of swimming pool water," and the "pH of soil." In what other situations have you heard the term "pH" used before? What do you think the term means? Discuss your ideas with your group.

In your home, as well as in the laboratory, there are solutions we now know to be acids or bases. We have not yet shown *how acidic* or *how basic* they are. In many situations, it is important to know how acidic or basic something is. Farmers, for example, may need to know how acidic or basic their soil is, so that they can prepare the soil for the planting of crops. There is a special kind of indicator paper called **universal indicator paper**, which turns a different colour depending on how acidic or basic a solution is. Each colour can be matched up with a numerical value on the pH scale.

Materials

Bring in some everyday solutions that you know are either acids or bases. Remember that many cleaning liquids are basic and many foods are acidic. Your teacher will supply some laboratory solutions for testing. These solutions will be buffered to different degrees of acidity or basicity.

Make a list of the solutions and equipment you will need, including the universal indicator paper. Ask your teacher to approve your list.

Method

1. Plan how you will test the buffered solutions and your everyday solutions with universal indicator paper. Show your plan to your teacher before you begin.

2. Record your observations in a data chart, such as the one below.

Solution	Colour of indicator	pH of solution

A solution is more acidic the *lower* its pH.

A solution is more basic the *higher* its pH.

Follow-up

1. Basic solutions have pH values above 7. Acidic solutions have pH values below 7. The pH of a neutral solution is 7. Use the information in Figure 3 to list the acidic solutions, beginning with the least acidic and ending with the most acidic.

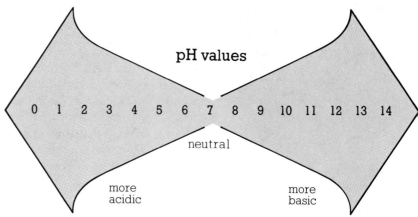

Fig. 3

2. Make a similar list of the basic solutions, beginning with the least basic.

3. (a) What sorts of consumer goods appear in each list?
   (b) Where do consumer goods fit in on the scale of acidity or basicity?

4. (a) Test the pH of the skin on your hands and your scalp with universal indicator paper.
   (b) For safety reasons, do not test the the inside of your mouth. Instead, ask your teacher for a typical pH value for this part of the body.

5. (a) What solutions should you avoid touching with your hands? Why?
   (b) What cleaning liquids would make irritating shampoos? Why?
   (c) What liquids would be safe to drink and which would be dangerous? Why?

6. Propose a definition for the term pH. Compare your proposal with those of your classmates. Decide on a definition that is agreeable to all the members of your class.

7. Write a report on your observations.

Hemoglobin is an iron-containing compound that carries oxygen through the bloodstream.

A base in water is called an **alkali**, hence the term "alkalosis."

Fig. 4 *The pH of blood is kept within a narrow range by buffers.*

BUFFERS PLAY AN IMPORTANT ROLE IN BLOOD

The normal pH of blood is in the range 7.35 to 7.45. If the pH of blood drops below 7.35, then a condition known as "acidosis" exists. If the pH of blood increases above 7.45, then "alkalosis" exits. If acidosis or alkalosis is not quickly remedied, death soon occurs.

Normally, blood stays within the safe pH range because of the presence of buffers. There are hydrogen carbonate buffers, phosphate buffers, and hemoglobin buffers. These buffers react with excess acid to protect the body from acidosis. The buffers also react with excess base to protect the body from alkalosis. In either case, the pH of the blood remains within safe limits.

Acidosis is most often caused by starvation, diabetes, kidney failure, and shock. In the event of shock, the normal oxygen supply to body cells is greatly reduced. As a result, the cells produce lactic acid, and the blood rapidly becomes acidic.

Alkalosis occurs less frequently. It may be caused by breathing too rapidly, using great amounts of laxatives, and as a side effect of pneumonia.

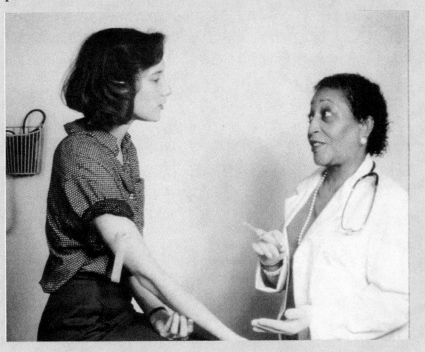

| Investigation | ## 12.3 The Turning Point of an Indicator |

You observed in investigation 12.1, that a number of indicators had different colours in acidic and basic solutions. In the present investigation, you will complete your report to the chemical manufacturing company by finding out more about the new indicators.

In investigation 12.1, what was the colour of each dye you tested in the sample acid solution? in the sample base solution? What do you think would happen if you changed the pH of the acid or base and then used it to test each dye? Do you think that all indicators change colour at the same pH? If so, what do you think their turning point is? If you think that different indicators have different turning points, can you predict what their turning points are? Justify any predictions you make and discuss them with your group.

The pH at which an indicator changes colour is called its **turning point**.

Materials

List the solutions you will need to determine the turning point of one of the indicators you used in investigation 12.1. Add the equipment you will need and get your teacher's approval of the list.

Don't forget your safety goggles!

Many indicators are toxic.

Handle acids and bases with care.

Method

1. Decide how to determine the colour of your indicator over a range of pH values. List the steps you will take and have your teacher check the list.

2. Prepare a data chart for your observations.

Follow-up

1. How does the indicator behave in solutions with different pH values?

2. (a) From your data chart, select the pH value at which your indicator changes colour.
   (b) Does the colour change occur all at once or does one colour gradually blend into the other?

3. Could you use the indicator you studied in this investigation to distinguish the following?
   (a) a solution of pH 4 and a solution of pH 7
   (b) a solution of pH 6 and a solution of pH 8
   (c) a solution of pH 7 and a solution of pH 9

Give reasons for each of your answers.

4. Record the results your classmates obtained with other indicators. Which indicator(s) could you use to distinguish each pair of solutions listed in question 3? Explain your choice for each pair of solutions.

5. How are colour indicators used in everyday life and in science?

| Investigation | ## 12.4 The Turning Point of a Mixture of Indicators |

Sometimes the turning point of an indicator is difficult to detect. For example, bromothymol blue changes from yellow to blue at a pH of about 7. As the yellow gradually turns to blue, a greenish shade is observed. It is quite difficult to see when yellow becomes green or when green becomes blue.

Now, the chemical manufacturer is getting nervous. The company's testing laboratory suggests you mix a pair of indicators to see if you can make the colour change more distinct.

Consider the mixtures of indicators suggested by your teacher. On the basis of the behaviour of the two indicators in the mixture, predict the turning point of the mixture. What colour do you predict for the mixture in acid solution? in base solution? How did you decide what the colours would be?

Don't forget your safety goggles!

Many indicators are toxic.

Handle acids and bases with care.

Materials

Get a sample of a mixed indicator from your teacher. What other solutions and equipment will you need? Present a list of materials to your teacher for approval.

Method

1. Plan a procedure that will determine if your predictions were correct.

2. Have the procedure checked by your teacher and carry it out.

Follow-up

1. (a) What turning point did you predict for the mixture?
   (b) What was the observed turning point of the mixture?
   (c) Why was there a difference?

2. Compare the pH range over which the colour of the mixture changed to the ranges for the individual indicators in the mixture.

3. In investigation 12.2, you used universal indicator. Can you tell from your results in investigations 12.3 and 12.4 whether universal indicator is a single indicator or a mixture of indicators? Explain.

4. Prepare a report to the chemical manufacturer on the properties of the indicators sent to you for testing.

Investigation	## 12.5 The Turning Point of a Natural Indicator

In 1684, Robert Boyle observed that basic solutions are slippery and that they can restore the colours of vegetable dyes that have been changed by acids. Boyle's observation was the first reference to chemical indicators. Many highly coloured vegetable and fruit juices make excellent indicators.

Fig. 5 *Many vegetables contain natural indicators.*

Materials

Select a natural indicator from among the following: carrot juice, tea, red cabbage juice, grape juice, blueberry juice, beet juice, and cranberry juice. If the indicator has not been prepared for you, boil the fruit or vegetable in just enough water to cover it. Continue boiling until the water takes on the colour of the fruit or vegetable. Allow the liquid to cool and pour it into dropper bottles.

What other solutions and materials do you need to determine the turning point? Also, include whatever you need to answer follow-up question 4. Ask your teacher to look over your list before you continue.

Method

1. Apply the same method you used to determine the turning points of the other chemical indicators. Follow all necessary safety precautions.

2. Record your own results and those of your classmates.

Follow-up

1. At what pH did your indicator change colour?

2. If you had only the natural indicators used in this investigation, which one would you use to tell the difference between the following?
   (a) a solution of pH 4 and a solution of pH 7
   (b) a solution of pH 6 and a solution of pH 8
   (c) a solution of pH 7 and a solution of pH 9

   Give reasons for each of your choices.

3. If you could use only one natural indicator to help you decide the pH of a variety of different solutions, which one would you choose? Explain your choice.

4. By demonstrating the effects of acids and bases on several vegetable dyes, show how Robert Boyle contributed to the study of indicators.

5. Write a report on natural indicators that you and your classmates have studied.

| Investigation | 12.6 pH: A Matter of Concentration |

In 1909, Soren Sorenson, a Danish biochemist, was trying to control the amount of acid in beer. He developed the pH scale to describe the degree of acidity.

Handle acids and bases with care.

We have seen how the pH of a solution is related to the amount of acid or base dissolved in it. Perhaps you have noticed that, while it can tell us how acidic or basic a solution is, pH itself has no units.

Since all acid solutions contain H^{1+} ions, the pH must depend on the concentration of H^{1+} ions in an acid solution. As the pH *increases*, what happens to the concentration of H^{1+}?

Since all base solutions contain OH^{1-} ions, then the pH of a base solution must depend on the OH^{1-} ion concentration. As the pH *increases*, what happens to the OH^{1-} concentration?

Discuss your answers to the above questions with your group. Then test your hypothesis as follows.

Materials

You will be supplied with six solutions. They contain 0.01 mol/L, 0.001 mol/L, and 0.0001 mol/L H^{1+} ions, and 0.01 mol/L, 0.001 mol/L, and 0.0001 mol/L OH^{1-} ions. List everything you will need to test the pH of these solutions. Have your list approved by your teacher.

Method

1. List the steps in your testing procedure. Ask your teacher to approve your list.

2. Record your data in a chart.

Follow-up

1. (a) Does a higher pH correspond to a higher or lower concentration of H^{1+} ions?
   (b) Does you answer to (a) agree with your earlier hypothesis?

2. (a) Does a higher pH correspond to a higher or lower concentration of OH^{1-} ions?
   (b) Does your answer to (a) agree with your earlier hypothesis?

3. As the concentration of H^{1+} or OH^{1-} ions changes by a factor of 10, by how many units does the pH change?

4. Most pH values lie in the range of 0 to 14. The pH of a neutral solution is 7. As a solution becomes more acidic or more basic, does its pH move closer to 7 or further away from 7?

5. Consider the acids with pH values of 2 and 3.
   (a) Which acid has the greater concentration of H^{1+} ions?
   (b) By what factor do the H^{1+} ion concentrations differ?

6. Consider two bases with pH values of 8 and 10.
   (a) Which base has the greater concentration of OH^{1-} ions?
   (b) By what factor do the OH^{1-} ion concentrations differ?

7. (a) What is the pH of a solution that is 1000 times more acidic than a solution of pH 5?
   (b) What is the pH of a solution that is 10 times less basic than a solution of pH 12?

8. The pH of a solution may also be called its "Sorenson Index." Why?

9. (a) Examine the pH values of the solutions that contain 0.01 mol/L, 0.001 mol/L, and 0.0001 mol/L H^{1+} ions. Can you find any mathematical relationship between the H^{1+} ion concentrations and the pH values? (*Hint*: Consider the concentrations written in scientific notation as 1×10^{-2} mol/L, 1×10^{-3} mol/L, and 1×10^{-4} mol/L.)
   (b) On the basis of your answer to (a), what do you think the H^{1+} ion concentration is for each of the following?
      (i) a solution of pH 6
      (ii) a solution of pH 9

| *Investigation* | *12.7 pH of Pure Water* |

If you rearrange the formula of water, H_2O, into the form HOH, you can see that water contains both the "H" found in acids and the "OH" found in bases. We know that when acid solutions, containing H^{1+} ions, combine with base solutions, containing OH^{1-} ions, they form water. Because hydrogen and oxygen are both nonmetals, water is a covalent compound. The proportion of free H^{1+} and OH^{1-} ions in a sample of pure water is very small. Let's now examine what the concentrations of these ions are in pure water.

Materials

pure water
universal indicator paper

Method

1. Use universal indicator paper to measure the pH of pure water.

2. Compare your result with those of your classmates.

Follow-up

1. In investigation 12.6, follow-up question 9, you related pH to the H^{1+} ion concentration expressed in mol/L. Now use the pH of pure water to determine the H^{1+} ion concentration in pure water.

2. Examine the molecular formula of water. Develop a hypothesis to describe how the concentrations of H^{1+} ions and OH^{1-} ions compare in pure water. Discuss your hypothesis with your group.

3. On the basis of your answers to questions 1 and 2, what is the concentration, in mol/L, of OH^{1-} ions in pure water?

4. We think of pure water as a neutral substance. Use your answers to questions 2 and 3 to define neutrality in terms of the concentrations of H^{1+} and OH^{1-} ions.

5. (a) What is the pH of a neutral solution, for example, a solution of table salt, NaCl?
   (b) Does your definition of neutrality from question 4 apply to neutral solutions, as well as to pure water?

| *Investigation* | ## 12.8 The Accurate Measurement of pH |

An environmental protection group has commissioned your class to come up with a reliable and accurate method for measuring pH. They have sent a sample of water from a lake in Québec for you to use in comparing methods.

Materials

Don't forget your safety goggles!

Assemble all the materials you have used to determine the pH of a solution. Strips of pH paper covering various pH ranges will be available to you. Your aim is to compare the consistency (precision) and accuracy of pH determinations for a series of acid and base solutions. If a pH meter (Figure 6) is available, you can use it to test the same solutions. Make sure that your teacher approves your list before you continue.

To standardize a pH meter, we use a buffer solution of accurately known pH. The meter is adjusted so that the pH reading equals the pH of the solution.

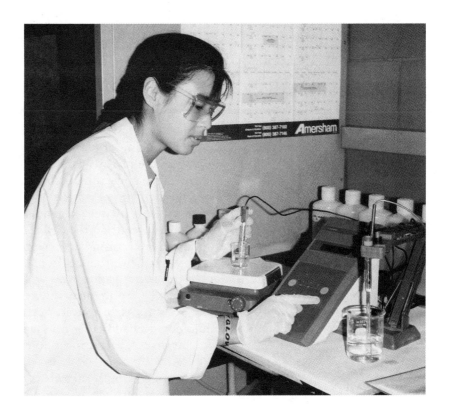

Fig. 6 *A pH meter is an expensive instrument used to measure pH in industry. If you use one, follow the correct procedure shown by your teacher.*

Method

1. Outline your procedure for comparing the various ways of measuring pH. Ask your teacher to check your procedure before you begin testing.

2. Compare your pH measurements with those of the other students in your class.

Follow-up

1. Which tests gave the most precise, that is, the most consistent, results?

2. (a) Which tests do you think were most accurate?
   (b) Check your answer to (a) by obtaining the correct pH values from your teacher.

3. How would you modify the technique for measuring pH in order to improve the precision and accuracy of the results?

4. Use your improved procedure to determine the pH of the sample of lake water.

| Investigation | 12.9 How pH is Used in Industry |

Fig. 7 *"Tears free" baby shampoos have the same pH as a baby's tears, and so these shampoos do not sting if they get into eyes.*

If a pH meter (Figure 6) is available, you can test almost any consumer product. If a meter is not available, you can use universal indicator solution or wide-range pH paper to test only products that are not highly coloured.

Have you ever bought a milk product that tasted sour? or ruined clothes by washing them? or used a product that irritated your skin? or made a chocolate cake that was too pale? Each of these problems might be traced to an incorrect pH level.

The checking of pH levels is important in many industries. For example, the pH of shampoos has to be carefully controlled (Figure 7). In the manufacture of chocolate cake mixes, baking soda (sodium hydrogen carbonate) is added to increase the pH if a dark brown cake is desired. The rate of growth of bacteria in cosmetics and food depends on the pH. The pH of some products is tested during manufacturing and before shipping. Some companies buy samples of their products from stores and then test their pH.

In this investigation, you will test the pH of some consumer products. There are many interesting groups of consumer products that you may wish to test. You could test the pH of the following: milk powders and other milk products; candies that dissolve (such as fruit drops); soft drinks; different types of flour (bleached, unbleached, pastry, whole wheat, buckwheat, corn, or rice flour); cocoa powders; juice crystals; baked goods (cookies, cakes, breads, muffins); mixes such as cake, muffin, bread, and pancake; personal care products such as shampoos, hair conditioners, and hair removers; detergents.

Materials

Select which consumer products you will test and plan how you will test them, either with a pH meter or with universal indicator paper or solution. When you have a complete list of materials and equipment, ask your teacher to check it.

Method

1. Dissolve the consumer product in water so that you can measure its pH. (Adding about 10 g of product to about 100 mL of water is usually appropriate.)

2. You may have to crush your product and stir it well in order to get it to dissolve in water.

3. If a product does not dissolve readily, let any undissolved solid material settle to the bottom and then pour off some of the clear liquid into a clean beaker for testing.

4. Plan your tests and have your plan approved by your teacher before you begin.

5. Record your observations in a chart. Be sure to include the brand name along with the type of product you are testing.

Follow-up

1. (a) Classify the consumer products you tested as acidic, basic, or neutral.
   (b) Were you surprised by any of the results? If so, which ones and why?

2. Why is a pH meter required to measure the pH of a highly coloured product?

3. Why are many industries concerned about the pH of their products?

Investigation

12.10 Analyzing a Household Base

Ammonium hydroxide is more correctly known as "aqueous ammonia." The term "ammonium hydroxide" is used here to emphasize that ammonia forms a base in water.

Many household cleaners contain ammonia. Ammonia gas dissolves in water to form ammonium hydroxide solution, which helps in the removal of greases and oils. Because ammonium hydroxide is a base, it may be neutralized by an acid. It is neutralized by hydrochloric acid according to the following word equation:

ammonium hydroxide + hydrochloric acid →

ammonium chloride + water

A household cleaner should have enough ammonia in it to be effective as a grease-cutter but not enough to damage the delicate membranes in the nose and mouth of the person using it. In this investigation, we will determine exactly how much ammonia there is in a sample of a household cleaner.

Quantitative analysis is the determination of the amounts of substances in a given material. Titration is a method of quantitative analysis that compares the volumes of two reacting solutions.

The procedure chemists use to determine the concentration of solutions is called a **titration**. It is essentially the same procedure you followed in investigation 9.8, when you neutralized sodium hydroxide solution by slowly adding hydrochloric acid solution to it drop by drop. The main difference in a titration is that chemists carry out the neutralization using a burette to deliver the solution. As with the earlier technique, the titration continues until the colour of an indicator *just* changes. The colour change is known as the **end point** of a titration.

Fig. 8 *The correct use of a burette.*

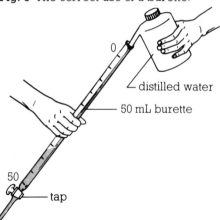

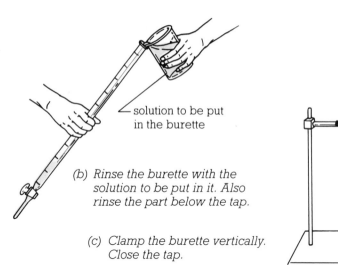

(a) *Rinse the burette with distilled water. Open the tap to let the water run out. This rinses the part below the tap.*

0

distilled water

50 mL burette

50

tap

(b) *Rinse the burette with the solution to be put in it. Also rinse the part below the tap.*

solution to be put in the burette

(c) *Clamp the burette vertically. Close the tap.*

funnel

(d) *Use a funnel to fill the burette to above the zero mark. Open the tap to allow some solution to run out. Close the tap again.*

meniscus

0

(e) *Add or run out more solution until the bottom of the meniscus is at the zero mark (or any other mark).*

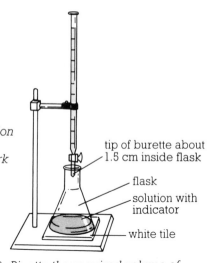

tip of burette about 1.5 cm inside flask

flask

solution with indicator

white tile

(f) *Pipette the required volume of another solution into a clean flask. Place the flask on a white tile under the tip of the burette. Add indicator into the flask.*

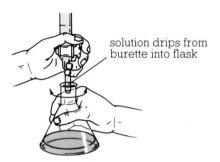

solution drips from burette into flask

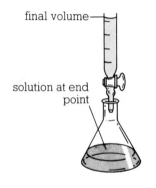

final volume

solution at end point

(g) *Control the tap by placing one hand around it from the rear. Use the other hand to swirl the flask as the solution in the burette runs down.*

(h) *When the end point is reached, close the tap immediately. Record the final volume of the solution in the burette.*

Materials

Your teacher will show you the sort of equipment a chemist uses in a titration (Figure 8). Make a list of this equipment and the purpose of each item. What other materials will you need? What chemicals will you need besides the sample of household cleaner? What information will you need to know about the acid you will use?

Make a "shopping list" of chemicals and equipment. Get your teacher's approval before you begin experimental work.

Method

1. Think about how you will analyze the household cleaner. You do not want to use up the entire bottle of cleaner, so you should plan on taking only a sample of it for analysis.

2. Ask your teacher to suggest an appropriate indicator.

3. Write out a plan for determining the volume of acid that will exactly neutralize all the ammonium hydroxide in your sample of household cleaner.

4. Ask your teacher to go over the plan with you before you begin experimental work.

Follow-up

1. (a) What volume of the household cleaner did you use?
   (b) How did you measure this volume in the most accurate way?

2. (a) Which indicator did you use?
   (b) What was the colour change you observed?

3. What volume of acid was needed to neutralize the household cleaner?

4. (a) Which of the solutions was more concentrated — the acid you used or the household cleaner?
   (b) How could you tell which one was more concentrated?

5. Why is a quick change of indicator colour important in titration?

6. (a) Now you are ready to calculate the concentration of the ammonium hydroxide in the household cleaner. Below is an equation you can use to calculate this concentration from the concentration of the acid you used.

For hydrochloric acid and nitric acid:

$$\frac{\text{concentration of base}}{(\text{mol/L})} = \frac{\genfrac{}{}{0pt}{}{\text{concentration of acid}}{(\text{mol/L})} \times \genfrac{}{}{0pt}{}{\text{volume of acid}}{(\text{mL})}}{\text{volume of base (mL)}}$$

For sulfuric acid:

$$\frac{\text{concentration of base}}{(\text{mol/L})} = \frac{2 \times \genfrac{}{}{0pt}{}{\text{concentration of acid}}{(\text{mol/L})} \times \genfrac{}{}{0pt}{}{\text{volume of acid}}{(\text{mL})}}{\text{volume of base (mL)}}$$

(b) Compare your calculated concentration with the values obtained by your classmates.

7. Write a report on your analysis of household ammonia.

| *Investigation* | ## 12.11 *Analyzing a Household Acid* |

Just as you can use an acid of known concentration to determine the concentration of a household base, you can use a base of known concentration to determine the concentration of a household acid. Vinegar is a solution of acetic acid. The word equation for the neutralization of acetic acid by the base sodium hydroxide is:

acetic acid + sodium hydroxide → sodium acetate + water

The label on a bottle of vinegar makes the claim that it contains 5% acetic acid by volume. The companies that bottle vinegar check their products regularly to be sure they are selling the strength of vinegar that they claim. Wise consumers want to know if they are getting their money's worth.

In this investigation, you will use the techniques you learned in the analysis of household ammonia to determine the concentration of acetic acid in vinegar.

Materials

Don't forget your safety goggles!

Refer to the list of equipment required for a titration and list what you will need to analyze vinegar. What sort of information about the chosen base solution will you need? Your teacher will check your list and may make some suggestions for improvement.

Method

1. You have already performed one titration, so you understand the way it is done. List the steps in your procedure.

2. Ask your teacher to check and approve your list before you begin.

Follow-up

1. What volume of vinegar did you use?

2. What volume of base did you need to neutralize the vinegar?

3. Calculate the concentration of acetic acid in vinegar with this equation.

$$\text{concentration of acid (mol/L)} = \frac{\text{concentration of base (mol/L)} \times \text{volume of base (mL)}}{\text{volume of acid (mL)}}$$

5.0% acetic acid solution, by volume, is 0.83 mol/L.

4. How does your sample of vinegar measure up? Is it at least 5.0% acetic acid?

5. What should the intelligent consumer do when a product does not live up to its advertisements?

| *Investigation* | ## 12.12 Studying the Reaction of an Acid With a Carbonate |

In investigation 9.6, we reacted hydrochloric acid with calcium carbonate. Do you remember how lively the reaction was? There was plenty of fizzing, as long as you had enough acid to react with the carbonate salt. Was the acid destroyed by reacting it with the carbonate? How could you prove that? How is this reaction similar to the reaction of a base, such as sodium hydroxide, with hydrochloric acid? Could the reaction of the acid with carbonate compounds be used to neutralize the acid? What gas is produced in the reaction of an acid with a carbonate? How could you identify the gas? Discuss your hypotheses with your group.

Water-soluble antacid tablets and powders contain both a carbonate salt and a solid acid. When you put these antacids in water, the carbonate and the acid react to produce lots of gas.

Materials

Don't forget your safety goggles!

You will need an acid solution and a compound containing the carbonate ion. For the latter, you might choose marble chips, which are made of calcium carbonate. What other possibilities are there?

You will need a way of knowing whether there is acid left in the solution at the end of the reaction. What method would you suggest?

You should also find out what gas is escaping during the reaction. What gas do you expect? What is the test for this gas?

Do you need to collect the gas formed so that you can identify it, or can you test the bubbles without collecting them? (Investigations 1.10 and 2.4 may give you some ideas.) Talk over with your teacher the kind of equipment you must build in order to test the gas produced. List everything you will need for the experiment and have the list approved.

Method

1. Prepare a list of directions for yourself and draw a picture of the apparatus you plan to use.

2. Get your teacher's approval, and begin your work.

3. Be sure to write down all of your observations.

Follow-up

1. (a) What is the name and the formula of the gas that is formed in the reaction?
   (b) How did you decide it was this gas and not another?

2. (a) Was there any acid left when the bubbling stopped?
   (b) What experimental observation helped you answer that question?

3. Copy the following word equation into your notebook. Complete the word equation so that it represents the reaction between calcium carbonate and hydrochloric acid:

   calcium carbonate + hydrochloric acid $\rightarrow$

   calcium chloride + water +

4. (a) Propose another way of representing the chemical change you have just observed.
   (b) Compare your representation with those of other students. Discuss the strengths and weaknesses of each representation.

5. Acid rain ruins our lakes and rivers. Plants and fish cannot live in lakes and rivers that are too acidic. Can you suggest a way of neutralizing the acid in our lakes, so that the environment is not harmed even more?

6. In your grandparents' childhood, people took baking soda (sodium hydrogen carbonate) to relieve an overly acidic stomach. Explain why this was, and still is, a good remedy.

Fig. 9 *Antacids contain bases to neutralize stomach acid.*

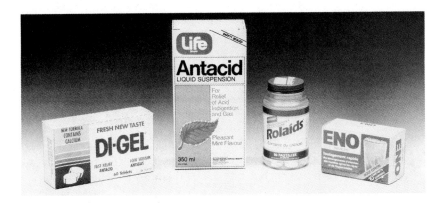

Investigation	**12.13 Keeping Track of Reagents and Products**

If we *make* products in a chemical reaction, is there *more* matter present after the reaction than before? Does the matter weigh more after the reaction than before? If we take another look at a familiar type of reaction, such as neutralization, perhaps we can discover the answers to these questions. Suggest what you think the answers will be and discuss your ideas with your group.

Materials

Don't forget your safety goggles!

Study a neutralization reaction between two solutions. You will obviously need an acid and a base. If you want to measure the mass, you will need a balance. What precautions must you take so that no material is lost through spilling or spattering during your reaction? List the chemicals and equipment you will need, and get your teacher's approval before you go on.

Handle acids and bases with care.

Method

1. Think about the measurements you must make in order to find out if the mass increases, decreases, or stays the same during a chemical reaction. Keep in mind the order in which you make the measurements. How will you know when the reaction is complete?

2. Write down your planned procedure and have your teacher check it.

3. Carry out the procedure and record all your results in a data table.

Follow-up

1. (a) Did the mass increase, decrease, or stay the same in your reaction?
   (b) Compare your answer to (a) with the findings of other students. What is the general conclusion of the class?

2. Were you surprised by the results of your investigation? Explain.

3. Why is it important to avoid any spillage when you add one reagent to another?

4. (a) Do you think you would get the same results if you repeated this experiment using the reaction of the acid with calcium carbonate?
   (b) What special problems are there in measuring mass changes in this reaction?
   (c) What changes could you make to your equipment to overcome the problems described in (b).

5. (a) Suggest a chemical law that would explain the relationship between the mass of the reactants and the mass of the products in a chemical reaction.
   (b) Compare your suggested law with those of other students.

| Investigation | ## 12.14 Determining the Product of a Neutralization |

We know what the reagents are in the neutralization carried out in investigation 12.13, but we have not yet determined the nature of the product. If the water that dissolved the reactants and now dissolves the product can be boiled away, the identity of the product should be revealed.

Materials

Place a wire gauze on top of the beaker to protect against the spattering of hot liquids. Stop heating before the contents of the beaker are completely dry.

List the chemicals and the equipment you will need (refer to Figure 10). Remember that you will have to determine when you have exactly neutralized the acid solution with the base solution. You want to identify the product, so choose a way of indicating neutrality that will not change the product's appearance.

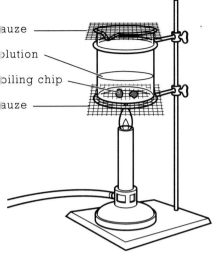

auze

olution

oiling chip

auze

Fig. 10 *Gently boil the solution to remove the water.*

Method

1. List all the steps you intend to take in order to remove the water from the solution of product.

2. Have your teacher check and approve your list before you begin.

Follow-up

1. (a) Can you identify from its appearance the solid compound formed in the neutralization of the acid and base?
   (b) Was any other product formed? If so, what was it?

2. (a) Write the word equation for the reaction between the acid and base.
   (b) Replace the words with the correct chemical formulas of the reactants and products in the neutralization.

3. (a) Have any atoms been created or destroyed in this chemical reaction?
   (b) How do you know?

4. (a) Is the solid product of the neutralization reaction an acid, a base, or a salt?
   (b) Use your answer to (a) and your answer to question 1(b) to complete the following general word equation for the reaction of any acid with any base:

$$ACID + BASE \rightarrow \quad + $$

12.15 The Law of Conservation of Mass

Your result in investigation 12.13 showed that mass neither increases nor decreases during a chemical reaction. This finding is known as the **law of conservation of mass**. For example, in the reaction:

$$hydrogen + oxygen \rightarrow water$$

the total mass of the hydrogen and the oxygen that react equals the mass of the water produced.

Thus, in any chemical reaction, we cannot lose or gain atoms. The total number of each type of atom must be the same in the reactants as in the products. To understand this further, let us write the formula of each chemical in the above word equation:

$$hydrogen + oxygen \rightarrow water$$
$$H_2 \quad + \quad O_2 \quad \rightarrow H_2O$$

We can also show this in pictures:

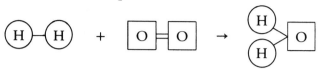

On the reactant side, we have two hydrogen atoms and two oxygen atoms. On the product side, we have two hydrogen atoms but only one oxygen atom. The number of atoms of each type is not balanced, and the equation below is called a **skeleton equation.**

$$H_2 + O_2 \rightarrow H_2O$$

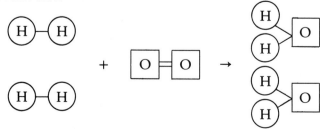

Since we cannot change the formula of water, the only way we can get two oxygen atoms on the product side is to add another water molecule.

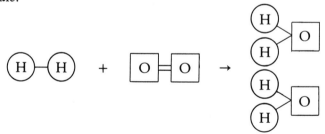

This step causes another problem. The product side now contains four hydrogen atoms and the reactant side only two hydrogen atoms. We must therefore include another hydrogen molecule on the reactant side.

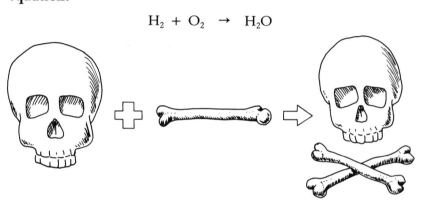

We have now balanced all the atoms that take part in this chemical reaction. All the atoms that we started with are still there, only rearranged into different molecules. The **balanced chemical equation** shows the correct numbers of all the atoms involved in the reaction. Of course, we don't use pictures to write it. Instead, we use symbols. In symbol form, the balanced chemical equation is:

$$2H_2 + O_2 \rightarrow 2H_2O$$

When balancing chemical equations, it is essential that you start with the correct formula of every reactant and product. Also, you must be able to count the total number of each type of atom in a particular formula. For example, magnesium phosphate, $Mg_3(PO_4)_2$, contains three atoms of magnesium, two atoms of phosphorus, and eight atoms of oxygen.

Chemically Speaking

Marie Paulze Lavoisier translated and annotated material that her husband required, illustrated his treatise on chemistry, and edited his memoirs. Mme Lavoisier discussed scientific theory with the major scientists of the time.

LAVOISIER'S WORK

The greatest contribution of the French scientist Antoine Laurent Lavoisier (1743-1794) to chemistry was the addition of quantitative measurements to what had been a simple cataloguing of observations. He and his wife, who was his collaborator, took painstakingly careful measurements. Their findings led to the proposal that, when a reaction takes place in a closed system, there is no increase or decrease in mass. We know this statement today as the law of conservation of mass.

Joseph Priestly, an Englishman, had observed that heating mercury(II) oxide caused it to turn into a silvery liquid, mercury, and to produce a colourless gas. However, it was only after Lavoisier performed this experiment in a closed system that this reaction could be identified as the decomposition of the compound into its elements, mercury and oxygen. Using Robert Boyle's definition of an element as a simple substance that could not be broken down further, Lavoisier identified and named several elements. He included oxygen and hydrogen in the table of elements of his *Elementary Treatise* on chemistry, considered the world's first chemistry textbook.

Fig. 11 *Antoine Lavoisier (1743–1794) made accurate measurements of mass. His work led to the law of conservation of mass.*

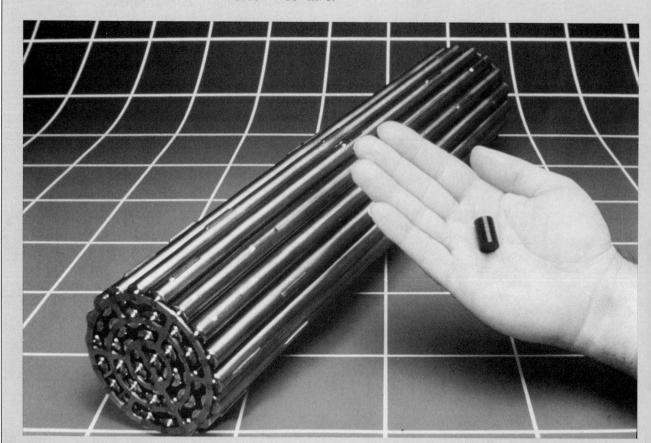

ALBERT EINSTEIN

Albert Einstein (1879–1955) will forever be remembered for proposing the idea that matter and energy are interconvertible. This idea means that matter can be turned into energy, and energy can be turned into matter. The relationship between the two is summarized in the famous equation $E = mc^2$. In this equation, the energy equivalent of a mass, in kilograms, is calculated by multiplying the mass by the speed of light squared. The speed of light is 3.00×10^8 m/s.

Fig. 12 *Nuclear energy comes from the controlled splitting of atomic nuclei.*

This kind of matter-to-energy conversion is observed in the splitting of atomic nuclei. It is the source of energy in nuclear reactors. Does the law of conservation of mass apply to nuclear changes? Can you suggest a way of modifying the law to include Einstein's findings?

Questions

1. For each of the following, write the formula and indicate the number of atoms of each type:
   (a) aluminum carbonate
   (b) calcium hydroxide
   (c) cadmium phosphate
   (d) magnesium hydrogen carbonate
   (e) hydrogen sulfate
   (f) sodium hydroxide
   (g) dinitrogen trioxide
   (h) lithium carbonate
   (i) barium nitrate
   (j) ammonium phosphate

2. Write an unbalanced chemical equation, that is, a skeleton equation, to represent each of the following reactions:
   (a) the synthesis of magnesium oxide from its elements
   (b) the decomposition of sodium chloride into its elements
   (c) the neutralization reaction of hydrochloric acid and calcium hydroxide

3. Joseph Proust, in 1799, proposed the law of constant composition. This law states that elements combine to form compounds in fixed mass ratios — for example, eight grams of oxygen combine with each gram of hydrogen to form water. Would Proust have been able to make this statement without the previous work of Lavoisier? Be able to defend your answer.

4. John Dalton stated that elements could combine in more than one fixed mass ratio. He cited the example of two compounds containing carbon and oxygen. In poisonous carbon monoxide, there are three grams of carbon for every four grams of oxygen. However, in carbon dioxide, there are three grams of carbon for every eight grams of oxygen. How did the work of Lavoisier contribute to this law of multiple proportions? Explain your answer.

12.16 Balancing Chemical Equations

Difficulty in balancing an equation often results from an incorrect chemical formula.

We balance chemical equations largely by trial and error. The usual steps are as follows:

1. Start with the word equation.

2. Convert the word equation into a skeleton equation by replacing each name with the correct formula.

3. If the equation contains a single element by itself, balance this type of atom at the end. The same rule applies when a particular element occurs in more than two substances in the equation.

4. In the skeleton equation, pick out the formula with the greatest number of atoms in it, and start working from there.

5. Balance atoms only by placing whole numbers in front of formulas. You *cannot* alter a formula.

6. When you have finished, check to see that you have used the lowest possible whole numbers. If not, divide all the numbers by the highest common factor.

As an example, follow the above steps in writing a balanced equation for the burning of propane, C_3H_8 (Figure 13).

Fig. 13 *Propane is the fuel used in gas barbeques and some vehicles.*

Remember which elements are written as molecules with two atoms each: H_2, N_2, O_2, F_2, Cl_2, Br_2, and I_2.

1. The word equation for the reaction is:
   propane + oxygen → carbon dioxide + water

2. The word equation is converted into a skeleton equation:
$$C_3H_8 + O_2 \rightarrow CO_2 + H_2O$$
   Remember oxygen gas has the formula O_2.

3. Here oxygen is a single element by itself, and it occurs in three of the substances in the equation. We will leave the balancing of oxygen until the end.

4. The formula C_3H_8 contains the most atoms. We will start our balancing using this formula.

5. The left-hand side of the equation contains three C atoms. To get three C atoms on the right-hand side, we place a "3" in front of the formula CO_2:

$$C_3H_8 + O_2 \rightarrow 3CO_2 + H_2O$$

The left-hand side of the equation contains eight H atoms. To get eight H atoms on the right-hand side, we place a "4" in front of the formula H_2O:

$$C_3H_8 + O_2 \rightarrow 3CO_2 + 4H_2O$$

Now, we can balance the oxygen atoms. The right-hand side of the equation contains a total of ten oxygen atoms (six from $3CO_2$ and four from $4H_2O$). To get ten oxygen atoms on the left-hand side, we must place a "5" in front of the formula for the oxygen molecule. The equation now becomes:

$$C_3H_8 + 5O_2 \rightarrow 3CO_2 + 4H_2O$$

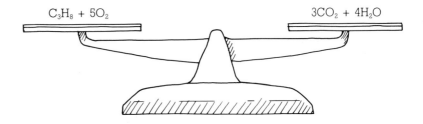

6. There is no common factor, other than 1, for the numbers 1, 5, 3, and 4, so our set of numbers represents the lowest possible whole numbers for the equation. The final equation represents the balanced chemical equation for the reaction.

Questions

1. Balance the following skeleton equations:
   (a) $Cu + O_2 \rightarrow CuO$
   (b) $Zn + HCl \rightarrow ZnCl_2 + H_2$
   (c) $P + H_2 \rightarrow PH_3$
   (d) $Al + CuSO_4 \rightarrow Cu + Al_2(SO_4)_3$
   (e) $H_2SO_4 + Al \rightarrow H_2 + Al_2(SO_4)_3$
   (f) $Na + H_2O \rightarrow H_2 + NaOH$
   (g) $NH_3 + O_2 \rightarrow NO_2 + H_2O$
   (h) $Na_2CO_3 + H_3PO_4 \rightarrow Na_3PO_4 + H_2O + CO_2$
   (i) $FeS + O_2 \rightarrow Fe_2O_3 + SO_2$
   (j) $Na_2CO_3 + FeCl_3 + H_2O \rightarrow Fe(OH)_3 + NaCl + CO_2$

The formulas of some common acids: sulfuric acid, H_2SO_4; nitric acid, HNO_3; phosphoric acid, H_3PO_4.

2. Write a balanced chemical equation given each of the following word equations:
   (a) sodium oxide + water $\rightarrow$ sodium hydroxide
   (b) sodium hydroxide + sulfuric acid $\rightarrow$
   sodium sulfate + water
   (c) sodium + oxygen $\rightarrow$ sodium oxide
   (d) sodium phosphate + silver nitrate $\rightarrow$
   silver phosphate + sodium nitrate
   (e) potassium carbonate + calcium chloride $\rightarrow$
   calcium carbonate + potassium chloride
   (f) barium nitrate + phosphoric acid $\rightarrow$
   barium phosphate + nitric acid
   (g) ammonium sulfate + calcium nitrate $\rightarrow$
   ammonium nitrate + calcium sulfate
   (h) barium hydroxide + phosphoric acid $\rightarrow$
   barium phosphate + water
   (i) magnesium carbonate + nitric acid $\rightarrow$
   magnesium nitrate + water + carbon dioxide
   (j) aluminum + oxygen $\rightarrow$ aluminum oxide

3. Balance the following skeleton equations:
   (a) $C_5H_{12} + O_2 \rightarrow CO_2 + H_2O$
   (b) $C_6H_6 + O_2 \rightarrow CO_2 + H_2O$
   (c) $K + H_2O \rightarrow H_2 + KOH$
   (d) $Mg(OH)_2 + H_3PO_4 \rightarrow Mg_3(PO_4)_2 + H_2O$
   (e) $KClO_3 \rightarrow KCl + O_2$
   (f) $MnO_2 + HCl \rightarrow MnCl_2 + Cl_2 + H_2O$

4. Write a balanced chemical equation given each of the following word equations:
   (a) sodium hydroxide + aluminum chloride $\rightarrow$
   aluminum hydroxide + sodium chloride
   (b) calcium carbonate + nitric acid $\rightarrow$
   calcium nitrate + carbon dioxide + water
   (c) dinitrogen pentoxide + water $\rightarrow$ nitric acid
   (d) diphosphorus pentoxide + water $\rightarrow$ phosphoric acid

12.17 Predicting the Quantity of Product

Let's look more closely at the reaction of hydrogen gas with oxygen gas to form water. The balanced chemical equation is:

$$2H_2 + O_2 \rightarrow 2H_2O$$

- How many MOLECULES of water are produced from the reaction of 2 MOLECULES of hydrogen gas (H_2) with 1 MOLECULE of oxygen gas (O_2)?

- How many MOLECULES of water are produced from the reaction of 6 MOLECULES of hydrogen gas (H_2) with 3 MOLECULES of oxygen gas (O_2)?

- How many MOLECULES of water are produced from the reaction of 2 DOZEN MOLECULES of hydrogen gas (H_2) with 1 DOZEN MOLECULES of oxygen gas (O_2)?

- How many MOLECULES of water are produced from the reaction of $2 \times 6.02 \times 10^{23}$ MOLECULES of hydrogen gas (H_2) with 6.02×10^{23} MOLECULES of oxygen gas (O_2)?

- How many MOLES of water molecules are produced from the reaction of 2 MOLES of hydrogen molecules with 1 MOLE of oxygen molecules?

- How many MOLES of water molecules are produced from the reaction of 8 MOLES of hydrogen molecules with 4 MOLES of oxygen molecules?

We can see from this examination that the numbers preceding the chemical formulas in a balanced chemical equation tell us two things:

(1) the relative numbers of reactant and product molecules that react and form in a chemical reaction
(2) the relative numbers of moles of reactant and product molecules that react and form in a chemical reaction

The balanced chemical equation tells us, for example, that it takes 2 mol of H_2 for every 1 mol of O_2 to produce 2 mol of H_2O.

$$2 \text{ mol } H_2 : 1 \text{ mol } O_2 : 2 \text{ mol } H_2O$$

To determine the number of moles of water produced by the reaction of 3.0 mol of oxygen gas with sufficient hydrogen gas, multiply the moles of oxygen by the $H_2O : O_2$ mole ratio.

$$3.0 \, \text{mol} \, O_2 \times \frac{2 \, \text{mol} \, H_2O}{1 \, \text{mol} \, O_2} = 6.0 \, \text{mol} \, H_2O$$

To calculate the proportional amount of product formed from the reaction of any amount of reactant, multiply the moles of the reactant by the appropriate mole ratio.

$$n_{\text{product}} = n_{\text{reactant}} \times \text{mole ratio}$$

In the mole ratio, the moles of the substance to *find* are always on top and the moles of the substance *given* are always on the bottom.

In a chemical reaction, we never recover the amount of product we expect. We always lose some of the product in the recovery process.

363

Fig. 14 *When we transfer a solid from one container to another, we always lose some of it.*

Questions

1. The balanced equation for the reaction of hydrogen and oxygen tells us that 2 mol of hydrogen gas combines with 1 mol of oxygen gas to form 2 mol of water:

$$2H_2 + O_2 \rightarrow 2H_2O$$

   (a) What is the molar mass of hydrogen gas (H_2)?
   (b) What is the mass, in grams, of 2 mol of hydrogen gas (H_2)?
   (c) What is the molar mass of oxygen gas (O_2)?
   (d) What is the mass, in grams, of 1 mol of oxygen gas (O_2)?
   (e) What is the sum of the masses of 2 mol of hydrogen gas (H_2) and 1 mol of oxygen gas (O_2)?
   (f) What is the molar mass of water?
   (g) What is the mass, in grams, of 2 mol of water?
   (h) Are your findings in (e) and (g) consistent with the law of conservation of mass? Explain.

2. (a) Write a balanced chemical equation for the reaction you performed in investigation 12.12 between an acid and a carbonate.
   (b) Determine the molar mass of each reactant and product.
   (c) Establish whether the balanced equation is consistent with the law of conservation of mass. Show your reasoning.

3. Do you think that all balanced chemical equations conform to the law of conservation of mass? Explain your reasoning.

Investigation

12.18 Predicting and Measuring the Quantity of Product

The calcium carbonate formed is not soluble in water, so it will drop out of the solution as it is formed. The calcium carbonate will be wet after it is separated from the rest of the mixture. Dry the calcium carbonate before you weigh it.

The chemical equation for the reaction between calcium chloride solution and sodium carbonate solution is:

$$CaCl_2 + Na_2CO_3 \rightarrow 2NaCl + CaCO_3$$

If you reacted 10 mL of a 1 mol/L solution of calcium chloride with 10 mL of a 1 mol/L solution of sodium carbonate, how many moles of each reactant would you be using? How many moles of calcium carbonate do you expect will form when these two solutions react? (Remember that concentration (mol/L) × volume (L) = moles of solute.)

Materials

List the chemicals and equipment you need. Include in your list a glass rod to assist in careful pouring (Figure 15). Since calcium carbonate does not dissolve in water, you will need equipment to separate the insoluble product from the rest of the mixture, as well as equipment to dry and weigh the product. Ask your teacher to approve your list.

Fig. 15 *A glass rod will help you to pour liquids without spilling them.*

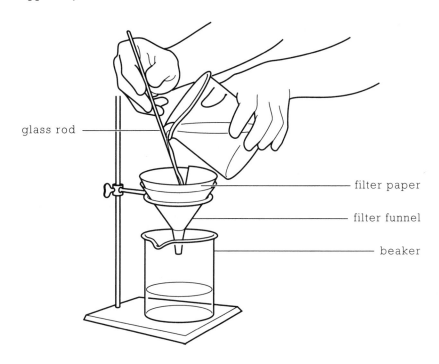

glass rod

filter paper

filter funnel

beaker

Method

1. List what you will do to determine if your predicted quantity of product was correct. Your list should contain directions on reacting the solutions, removing the precipitate, and drying and weighing it. When you make up your list, make sure you think about ways of collecting as much precipitate as possible. (You want to avoid losing very much of it.)

2. Ask your teacher to approve the list before you begin.

Follow-up

1. (a) How many moles of calcium carbonate did you predict would form?
   (b) What mass, in grams, of calcium carbonate did you expect?

2. (a) What mass, in grams, of calcium carbonate did you recover?
   (b) How close were you to the expected mass?

3. (a) Did you recover all the calcium carbonate that you made or did you lose some along the way?
   (b) At what point do you think you lost any of the product?

4. How are chemical engineers able to predict the mass of iron that they will produce from a tonne of iron ore?

12.19 Chemical Bookkeeping

Let's return to our study of the reaction between hydrogen gas and oxygen gas.

$$2H_2 + O_2 \rightarrow 2H_2O$$

We have learned that the numbers preceding the formulas indicate the relative numbers of moles of reactant and product in the chemical reaction.

$$2 \text{ mol } H_2 + 1 \text{ mol } O_2 \rightarrow 2 \text{ mol } H_2O$$

To calculate the mass of water produced when 0.200 mol of oxygen gas (O_2) is reacted with sufficient hydrogen, first use the water : oxygen mole ratio to determine the moles of water produced.

$$0.200 \text{ mol } O_2 \times \frac{2 \text{ mol } H_2O}{1 \text{ mol } O_2} = 0.400 \text{ mol } H_2O$$

When you know the number of moles of water produced, use the following relationship to calculate the mass of water produced.

$$n \text{ (mol)} = \frac{\text{mass of substance (g)}}{\text{molar mass of substance (g/mol)}}$$

$$n \times \text{molar mass of substance} = \text{mass of substance}$$

$$0.400 \text{ mol } H_2O \times \frac{18.0 \text{ g } H_2O}{1 \text{ mol } H_2O} = 7.20 \text{ g } H_2O$$

The calculation of the quantities of materials involved in chemical reactions is a branch of chemistry called **stoichiometry**.

The usual method for solving problems in chemical arithmetic is as follows:

Step 1: Write the balanced chemical equation.

Step 2: If necessary, calculate the number of moles of the substance *given*.

Step 3: Multiply the number of moles of the substance *given* by the mole ratio that relates the substance *to find* (on top) to the substance *given* (on the bottom). This calculation will give you the moles of the substance *to find*.

Step 4: Change the moles of the substance *to find* into the units required.

For example, what mass, in grams, of carbon dioxide is produced in the burning of 10.0 g of propane (C_3H_8)?

Step 1: $C_3H_8 + 5O_2 \rightarrow 3CO_2 + 4H_2O$

Step 2: Determine the number of moles of propane from the relationship:

$$n\,(\text{mol}) = \frac{\text{mass of substance (g)}}{\text{molar mass of substance (g/mol)}}$$

$$\text{Molar mass of } C_3H_8 = [3(12.0) + 8(1.0)]\,\text{g/mol}$$

$$= 44.0\,\text{g/mol}$$

$$n\,(C_3H_8) = \frac{10.0\,\text{g}\,C_3H_8}{44.0\,\text{g/mol}\,C_3H_8}$$

$$= 0.227\,\text{mol}\,C_3H_8$$

Step 3: Multiply the number of moles of propane by the mole ratio that relates propane to carbon dioxide.

$$0.227\,\text{mol}\,C_3H_8 \times \frac{3\,\text{mol}\,CO_2}{1\,\text{mol}\,C_3H_8} = 0.681\,\text{mol}\,CO_2$$

Step 4: Multiply the moles of carbon dioxide by the molar mass of carbon dioxide to determine the mass of carbon dioxide produced.

$$\text{Molar mass of } CO_2 = [12.0 + 2(16.0)]\,\text{g/mol}$$

$$= 44.0\,\text{g/mol}$$

$$0.681\,\text{mol}\,CO_2 \times 44.0\,\text{g/mol}\,CO_2 = 30.0\,\text{g}\,CO_2$$

As with all numerical calculations, it is a good idea to estimate the answer to make sure that your calculated answer is of the correct magnitude. The balanced chemical equation shows that, when 1 mol (44.0 g) of C_3H_8 reacts, 3 mol (132 g) of CO_2 should be produced. In the above example, we begin with 10.0 g of C_3H_8, which is a little less than one-quarter of 44.0 g. Therefore, we expect to make a little less than one-quarter of 132 g of CO_2. Dividing 132 g by 4 gives 33.0 g, so the calculated answer of 30.0 g looks reasonable.

Questions

1. Show that the law of conservation of mass is obeyed in the following chemical reactions.
   (a) $C + O_2 \rightarrow CO_2$
   (b) $KOH + HNO_3 \rightarrow KNO_3 + H_2O$
   (c) $2Na + Cl_2 \rightarrow 2NaCl$
   (d) $4Al + 3O_2 \rightarrow 2Al_2O_3$
   (e) $NaHCO_3 + HCl \rightarrow NaCl + H_2O + CO_2$
   (f) $Na_2CO_3 + 2HCl \rightarrow 2NaCl + H_2O + CO_2$

2. In Montréal, burning fallen leaves is not permitted because it adds to air pollution. What would you say to someone who claims that burning leaves does no harm because the leaves have all "disappeared" in the burning process?

3. What does the law of conservation of mass tell you about the disposal of wastes?

4. Complete the following chart for the neutralization of sulfuric acid with sodium hydroxide.

H_2SO_4 +	2 NaOH →	Na_2SO_4 +	$2H_2O$
98 g	80 g	142 g	36 g
490 g			
	160 g		
	24 g		
			7.2 g

5. Complete the following chart for the neutralization of hydrochloric acid with calcium carbonate.

$CaCO_3$ +	2HCl →	$CaCl_2$ +	H_2O +	CO_2
100 g				
	100 g			
		100 g		
			100 g	
				100 g

6. Calculate the mass, in grams, of calcium hydroxide needed to neutralize 9.8 g of sulfuric acid. Remember to balance the equation first.

$$Ca(OH)_2 + H_2SO_4 \rightarrow CaSO_4 + H_2O$$

7. (a) What mass, in grams, of calcium carbonate is needed to neutralize 100 g of nitric acid (hydrogen nitrate)? Remember to balance the equation first.

$$HNO_3 + CaCO_3 \rightarrow Ca(NO_3)_2 + H_2O + CO_2$$

   (b) What mass, in grams, of carbon dioxide is produced when 134 g of nitric acid is neutralized with calcium carbonate?

P O I N T S · T O · R E C A L L

- Most solutions have pH values in the range 0 to 14. A neutral solution has a pH of 7.
- The more acidic a solution is, the further its pH is below 7. The more basic a solution is, the further its pH is above 7.
- The pH of a solution can be measured with a pH meter, universal indicator solution, pH paper, or some natural indicators.
- The pH at which an indicator changes colour is called its turning point.
- The pH of a commercial product may affect the properties of the product. Therefore, the checking of pH levels is important in many industries.
- Buffers prevent large changes in pH.
- The pH of a solution is directly related to the concentration, in mol/L, of H^{1+} ions in the solution.

- Neutral solutions have equal concentrations of H^{1+} and OH^{1-} ions.
- A titration is an accurate way to determine the concentration of a solution.
- An unbalanced chemical equation is called a skeleton equation.
- We balance chemical equations by placing whole numbers in front of the formulas of the reactants and products.
- The law of conservation of mass states that the mass neither increases nor decreases in a chemical reaction.
- A balanced chemical equation shows the ratio of the numbers of moles of substances involved in a chemical reaction.
- An acid and a base react to form a salt and water.

R E V I E W • Q U E S T I O N S

1. A certain solution of sodium hydroxide has a pH of 10.75. What would be the pH of a sodium hydroxide solution that was ten times more concentrated?

2. The pH of a certain solution of hydrochloric acid is 2.49. If this acid solution was diluted by a factor of 100, what would be the pH of the more dilute solution?

3. When using an indicator to determine whether a solution is acidic or basic, why should you never use more than a few drops of the indicator?

4. Universal pH paper or universal indicator solution can be used to determine the pH of an acid or base solution across the entire pH range. How do you think these indicator solutions are made?

5. One of the most common pieces of equipment found in a quality control laboratory in industry is a pH meter. Why do you think this is so?

6. Why is the term "pH" so often used in the advertising of personal care products, such as hair shampoos, hair conditioners, and cosmetics?

7. What is the pH of a solution in which the H^{1+} ion concentration is 1×10^{-5} mol/L?

8. In solution A, the H^{1+} ion concentration is 1×10^{-3} mol/L. In solution B, the value is 1×10^{-2} mol/L. Which solution is more acidic?

9. What special precautions must you take to prove that the reaction between hydrochloric acid and calcium carbonate obeys the law of conservation of mass?

10. The rotting of a log is a chemical reaction, yet the log seems to lose mass until it crumbles away to nothing. How do you explain this?

11. Consider the reaction: $C + O_2 \rightarrow CO_2$. What mass of carbon must burn in 3.2 g of oxygen to produce 4.4 g of carbon dioxide?

12. Why must a chemical equation be balanced?

13. Balance the following skeleton equations:
    (a) $N_2 + H_2 \rightarrow NH_3$
    (b) $K + O_2 \rightarrow K_2O$
    (c) $NaOH + H_3PO_4 \rightarrow Na_3PO_4 + H_2O$
    (d) $Pb_3O_4 \rightarrow PbO + O_2$
    (e) $(NH_4)_3PO_4 + ZnCl_2 \rightarrow$
$$NH_4Cl + Zn_3(PO_4)_2$$

14. Write a balanced chemical equation given each of the following word equations:
    (a) barium nitrate + sodium sulfate $\rightarrow$
        barium sulfate + sodium nitrate
    (b) silver nitrate + ammonium sulfide $\rightarrow$
        silver sulfide + ammonium nitrate
    (c) octane (C_8H_{18}) + oxygen $\rightarrow$
        carbon dioxide + water
    (d) calcium hydroxide + hydrochloric acid $\rightarrow$
        calcium chloride + water
    (e) bromine + sodium hydroxide $\rightarrow$
        oxygen + sodium bromide + water

15. For each of the following neutralization reactions, provide:
    (a) a complete word equation
    (b) a balanced chemical equation
     (i) sodium hydroxide
                + hydrochloric acid $\rightarrow$
     (ii) barium hydroxide + nitric acid $\rightarrow$
    (iii) ammonium hydroxide
                + phosphoric acid $\rightarrow$
    (iv) aluminum hydroxide + sulfuric acid $\rightarrow$

16. Calculate the mass of sodium hydroxide that neutralizes 14.7 g of sulfuric acid. The equation for the reaction is:

$$2NaOH + H_2SO_4 \rightarrow Na_2SO_4 + 2H_2O$$

17. What mass, in grams, of calcium carbonate must react with hydrochloric acid in order to produce 100 g of carbon dioxide gas?

18. What mass, in grams, of silver carbonate can form when 12.6 g of sodium carbonate is added to a silver nitrate solution? The unbalanced chemical equation for this reaction is:

$$AgNO_3 + Na_2CO_3 \rightarrow NaNO_3 + Ag_2CO_3$$

19. How many molecules of ammonia are expected to form when 6.00 g of hydrogen gas (H_2) combine with sufficient nitrogen gas? The unbalanced chemical equation for this reaction is:

$$N_2 + H_2 \rightarrow NH_3$$

13

CHEMISTRY, THE ENVIRONMENT, AND SOCIETY

Air pollution is a serious problem.

CONTENTS

Investigation	13.1 A Breath of Fresh Air?

Close your eyes and imagine a place where you would go for some "fresh air." What is so special about the air in this place? What is in the air?

Now imagine a place where the air is not so pleasant. Where do you find such places? What is in the air that you do not like? How did it get there?

You may think of fresh air as "pure." In fact, there is no such thing as "pure air." Air is a solution with many components. Dry air contains about 80% nitrogen gas and about 20% oxygen gas. The quantity of water vapour in the air varies. The humidity is a measure of the water vapour content of the air. Where does the water vapour come from? There is also a small percentage of carbon dioxide gas in the air. Carbon dioxide is absorbed by plants, which use it, along with water and the energy of the sun, to make more plant material. Where does the carbon dioxide come from? What do plants give back in return for the carbon dioxide they use?

One of the principal problems facing modern society stems from the fact that the products of chemical reactions do not simply disappear when they are added to the air. Our daily lives, our comfort, and our economy currently depend on the burning or combustion of fuels—gasoline, heating oil, natural gas, and coal.

When a reaction with oxygen occurs rapidly, giving off lots of heat and light, it is called, **combustion** or burning.

373

The products of all this combustion may seem to vanish into the air, but they do not. Instead, they change the environment in which we live.

No doubt you are aware of the serious environmental problem of **acid precipitation**. Do you know how it is formed? Does combustion have anything to do with it? Discuss your knowledge of acid precipitation within your group, then complete the investigation to learn more about the products of combustion.

If you were to burn typical metals (magnesium, for example) and typical nonmetals (powdered sulfur or carbon chips), you could collect the fumes produced and dissolve them in water. Then you could test each solution for its acidic or basic properties. From your knowledge of the chemical formulas of acids and bases, can you predict whether the combustion of metals and nonmetals will produce acids or bases? Discuss your ideas with your group.

Materials

Select the elements you will study. Avoid testing hazardous metals, like sodium, and hazardous nonmetals, like phosphorus. You can avoid irritating your eyes and the inner lining of your nose if you burn each of the elements in a deflagrating spoon lined with aluminum foil and held inside a large glass container of oxygen (Figures 1 and 2). The container should be covered as much as possible with a glass plate (Figures 3 and 4).

Don't forget your safety goggles!

If you burn steel wool (which is mainly composed of iron), do not touch the glass of the jar with the hot steel.

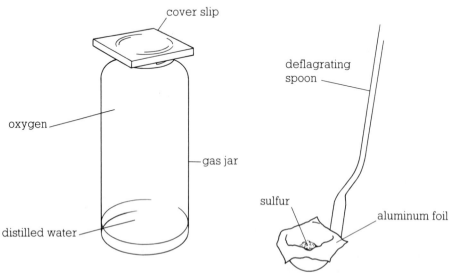

Fig. 1   Fig. 2

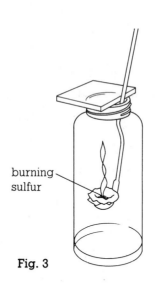

burning
sulfur

Fig. 3

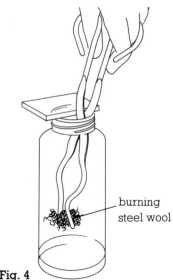

burning
steel wool

Fig. 4

You must not inhale the fumes.
Work in a fume hood or in a
well-ventilated room.

Do not stare directly at the
flame of burning magnesium.
Permanent eye damage could
result. Use cobalt blue glass.

Prepare a "shopping list" of materials you will need to burn the
selected metals and nonmetals, as well as to test their products in
water. Get your teacher's approval before continuing.

Method
1. List the steps you will take to burn the metals and nonmetals,
   collect the products, and test their acidic or basic properties in
   water.

2. Do not begin without your teacher's approval.

3. Present your observations in a data chart such as the one below.

Element burned			Observations of burning		Oxide in water tested with universal indicator		
Name and symbol	Metal or nonmetal	Name of oxide	Air	Pure oxygen	Colour	pH	Acidic/basic/neutral

Follow-up

1. What general name is given to the type of compound formed
   when an element reacts with oxygen?

2. When metal oxides react with water, what is formed? Copy the
   equation and enter the word *acid* or *base*.

   metal oxide + water →

375

3. When nonmetal oxides react with water, what type of substance is formed? Complete the general equation for this type of reaction.

nonmetal oxide + water $\rightarrow$

4. Fuels such as wood and gasoline contain compounds of carbon and hydrogen. When the fuels are burned, the carbon and hydrogen are changed into their oxides. Name the two oxides formed.

5. Copy and complete each word equation for a combustion reaction.
(a) _____ + oxygen $\rightarrow$ sulfur dioxide
(b) carbon + oxygen $\rightarrow$ _____
(c) iron + oxygen $\rightarrow$ _____
(d) magnesium + _____ $\rightarrow$ _____
(e) _____ + _____ $\rightarrow$ diphosphorus pentoxide
(f) aluminum + _____ $\rightarrow$ _____

13.2 Acid Rain and the Two Faces of Oxygen

Fig. 5 *The most acidic precipitation in North America generally falls on the eastern United States and Canada. Over much of Ontario, Québec, and the Atlantic Provinces, the average pH of precipitation is well below 5. Each numbered line on the map connects places where the average pH of precipitation is the same.*

Rain is naturally acidic, with a pH of about 5.6. Acid rain is more acidic and can have a pH as low as 2.0. In the worst cases, acid rain is about as acidic as vinegar.

Almost daily, we hear about acid precipitation. We hear that acid rain is destroying our buildings, lakes, crops, and forests. Even snow samples taken in the Far North are becoming more acidic. Based on your observations in investigation 13.1, where do you think the acids in precipitation come from? How do the acids get into the water cycle? Is there such a thing as "basic rain"? Give reasons for your answers and discuss them with your group.

Without oxygen, plant and animal life could not survive. But there are other, less beneficial effects of oxygen. Have you noticed that an apple or potato, once cut and exposed to air, goes brown? This is because of the reaction of oxygen with the foods. Some foods rot because of this reaction. Another negative effect of oxygen is its reaction with iron to form rust, called iron(III) oxide.

Many oxides are formed from combustion reactions, in which substances burn in oxygen. Combustion reactions are of great benefit to humans. For example, we use the burning of fuels to produce heat and other forms of energy. However, the burning of gasoline in automobiles accounts for 60% of the world's air pollution. Another 20% comes from burning fuels for home heating, and 20% from various industrial processes.

A number of pollutants released into the air can lead to acid precipitation, but the major sources are oxides of sulfur and nitrogen. In the atmosphere they are converted to sulfuric acid and nitric acid. Once the acids are dissolved in water, they can fall to the earth as rain and snow, or cover mountaintops with an acidic mist (Figure 6).

In mining districts of the Abitibi in Québec, valuable metals such as nickel, iron, and cobalt are taken from the earth in the form of crude sulfide ores. The metals are removed from the sulfur in the ores in a process called **roasting**. Close to 98% of the sulfur removed from the ores ends up as sulfur dioxide. The Abitibi is responsible for half the sulfur dioxide that contributes to the acid rain in Québec. To assist in reducing the harmful smokestack emissions by 50%, the Québec government has subsidized the building of a sulfuric acid plant.

Other industrial sources of sulfur dioxide emissions include petroleum refineries, pulp and paper plants, steel mills, and aluminum plants. Canada has set the year 1994 as the target date for reducing emissions from industrial sources by 50%.

Nitrogen oxides, which also produce acid rain, come mainly from running our automobiles. In 1984, Québec alone contributed 245 000 t of nitrogen oxides to the atmosphere.

You can slow the "browning" of some sliced fruits and vegetables by dipping the slices in an acidic solution, such as lemon juice. Try this when you next cut avocados, apples, bananas, pears, artichokes, or potatoes.

The term "iron(III)" means the compound contains Fe^{3+} ions.

Sulfuric acid, a very useful industrial chemical, is made by the **contact process**. In this process, the sulfur dioxide formed during the roasting of sulfide ores is converted to sulfur trioxide. The sulfur trioxide is then absorbed by a 98% sulfuric acid solution. Water is added to maintain the concentration at 98%. The concentrated sulfuric acid used in chemical laboratories is approximately 96%.

The letter t is the abbreviation for tonne. 1 t = 1000 kg

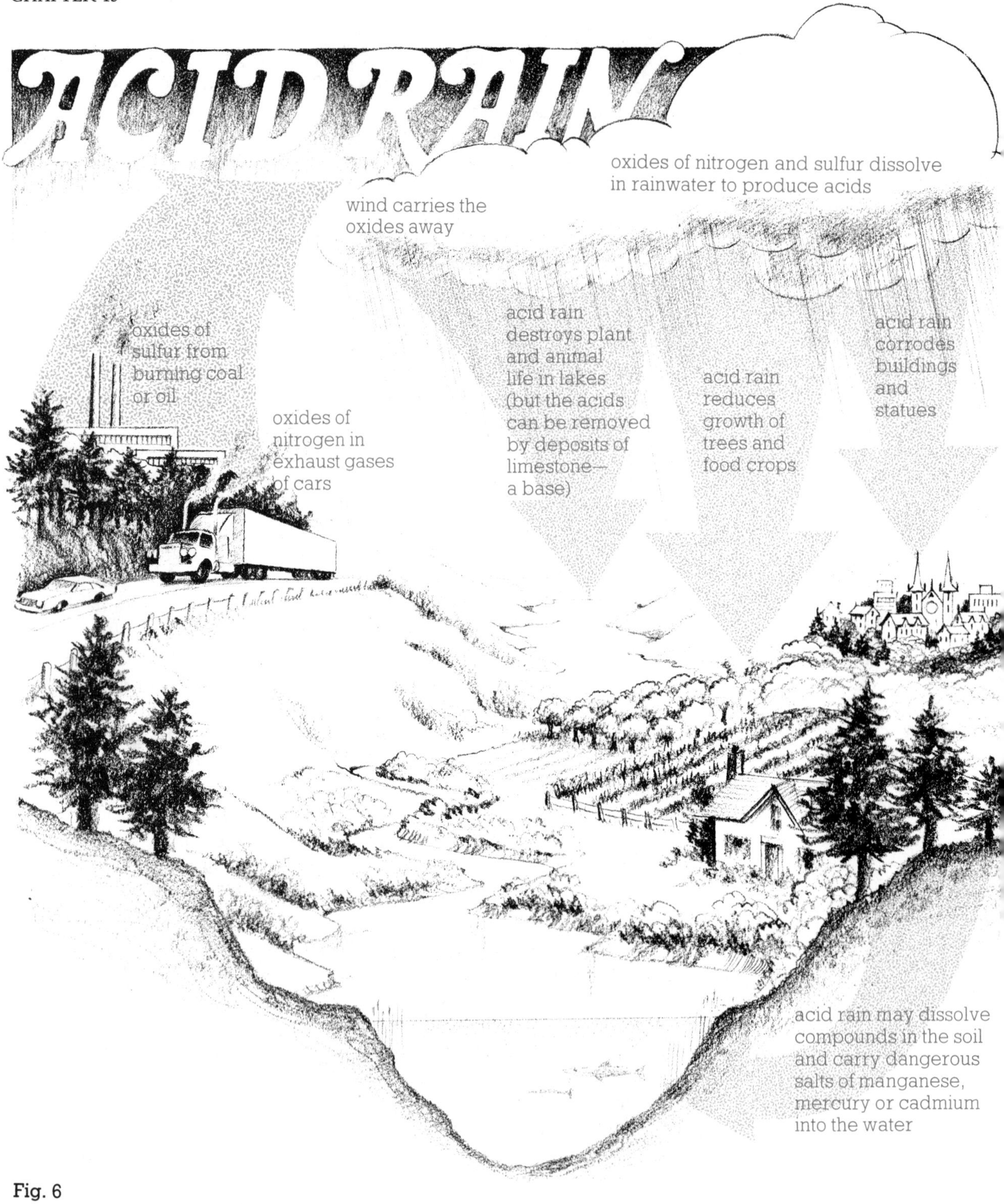

ACID RAIN

oxides of nitrogen and sulfur dissolve in rainwater to produce acids

wind carries the oxides away

oxides of sulfur from burning coal or oil

oxides of nitrogen in exhaust gases of cars

acid rain destroys plant and animal life in lakes (but the acids can be removed by deposits of limestone— a base)

acid rain reduces growth of trees and food crops

acid rain corrodes buildings and statues

acid rain may dissolve compounds in the soil and carry dangerous salts of manganese, mercury or cadmium into the water

Fig. 6

Questions

1. Nitrogen oxides are partially responsible for acid precipitation. Is this information consistent with your findings in investigation 13.1? Explain.

2. Draw on your everyday experiences to list as many positive effects and as many negative effects of combustion as you can think of.

3. Why is it important to reduce the quantity of acid air pollution across the continent, not just in your own region?

4. Briefly explain how acid precipitation is formed in the atmosphere.

5. Acid rain is often mentioned in the news. Why is acid rain considered so important?

6. Many major industries burn huge amounts of coal, which contains sulfur impurities. What can these industries do to reduce their emissions of oxides that cause acid rain?

Investigation	

13.3 The pH of Acid Precipitation

Just how acidic is the precipitation in your region? Is it as acidic as vinegar, or perhaps grapefruit juice or coffee? Make a guess, then test it in this investigation.

Materials

You will have to collect some rain or snow to test. What other solutions will you compare it to? List all the solutions and equipment you will need to test the pH of the precipitation, and add your comparison solutions to this list. Ask your teacher to check your list before you continue.

Do not forget your safety goggles!

Method

1. Plan how you will test the rain or snow, as well as the other solutions.

2. Outline all the steps and get your teacher's approval before starting.

Follow-up

1. How does the pH of the rain or snow compare to pH values for your other solutions?

2. What sort of materials will be affected by rain or snow of this pH?

3. Have you seen evidence of damage to objects in your area that could be caused by acid rain? If so, list the evidence.

13.4 Effects of Acid Rain on Our Environment

Fig. 7 *The shaded area shows the region of North America that produces the greatest quantity of acid pollutants. The arrows over the shaded area show major storm paths that carry the pollutants to other regions.*

Fig. 8 *Lakes in the Canadian Shield are very sensitive to acid rain.*

All of Québec and much of Ontario, the northern parts of the Prairie Provinces, and much of the Northwest Territories make up a geological region called the Canadian Shield. This area contains more than a million lakes, rivers, and streams that provide food, transportation, and water. The Canadian Shield is itself a very hard kind of rock, granite. There is no limestone (calcium carbonate), which is a softer rock that can neutralize acid precipitation. The Shield, which is so sensitive to acid rain, is in the direct path of westerly winds. These winds bring with them pollution from industries in central North America as well as the pollution Quebeckers produce.

The fish swimming in the lakes of the Canadian Shield, as well as the tiny plants and animals that fish eat, are in danger from the increasing acidity of the lakes. Acid rain causes the bones of the fish to weaken, making it difficult for them to swim. Aluminum, which is washed out of the soil by the acid rain, is poisonous to

The provincial government department of the environment in Québec reported in 1990 that, on account of the eight-fold increase in the average acidity of the lakes in south-western Québec in the past 60 years, some 10 000 populations of fish had been wiped out. Walleye, lake whitefish, and minnows are among the most sensitive species, while perch, northern pike, and lake trout are relatively resilient.

fish. Fish, in fact, begin to die even before the lake is considered unsafely acidic.

Farmers complain that acid rain is damaging their maple stands and other crops. In our cities, statues made of marble, which is essentially calcium carbonate, are being eaten away by acid rain, and metal and painted surfaces are being damaged.

Questions

1. Buildings, and marble and metal statues, suffer the effects of air pollution in cities. Explain these effects.

2. In 1986, the provincial ministry responsible for fish and game in Québec strongly advised hunters not to eat the livers and kidneys of moose, elk, and caribou killed in the province. These organs were found to contain high concentrations of toxic metals, such as cadmium. How could acid rain have played a role in making this meat unsafe to eat?

3. There is water on the surface of your skin. What type of substance can form on your skin if you live in or visit a city in which there are many cars? What effect can this type of substance have on metal jewellery or pearls worn next to the skin?

4. (a) What can you do as a citizen to reduce the amount of acid rain in your environment?
   (b) What organizations do you know about that are concerned with promoting a cleaner environment?

5. How would you propose to diminish the effect of acid precipitation on our biological or physical environment? Show the effectiveness of your proposal with a laboratory demonstration.

6. Research the steps that are already being taken to reduce acid emissions and to reduce the environmental impact of acid precipitation. Do you regard these steps as adequate? Explain.

7. In media reports of acid precipitation, is there agreement over its possible future effects? If so, what are these effects expected to be? If not, in what ways do the media reports differ?

8. Set up a debate in your class. The topic for the discussion is "Should the provincial government establish stricter guidelines for reducing acid rain emissions from local industries?" One team would argue for stricter government regulation and the other team would argue against it. Which side, do you think, has the stronger arguments?

13.5 Some Other Pollution Problems

Soaps and Detergents

Fig. 9 *Detergents contain additives that prevent scum formation.*

Personal cleanliness is very important in our society. That is probably why we have so many different types of detergents and soaps. Each product contains additives to:

- improve the whiteness of the clothing
- maintain the correct acidity level
- react with and prevent the minerals in the water from interfering in the cleaning process
- prevent corrosion of aluminum and porcelain surfaces
- act as perfumes and colouring agents to attract the consumer and mask any unpleasant odour

Phosphates added to detergents prevent scum formation (Figure 9). However, when waste water carries phosphates into rivers and lakes, the population of algae increases rapidly. These algae deplete the dissolved oxygen, which the animal life in the water needs for survival. To control this water pollution, governments have restricted the quantities of phosphates in detergents. In supermarkets, you will now find phosphate-free detergents. Their manufacturers have replaced phosphates with biodegradable substitutes.

A **biodegradable** substance is broken down into simpler substances by bacteria.

Plastics

In the 1930s, the era of synthetic fibres and plastics began. Two important polymers, nylon and Teflon, were discovered by researchers at the E. I. du Pont de Nemours Company in 1938.

Polymers are giant molecules made by bonding a large number of smaller molecules together. The word "polymer" comes from the Greek words *poly* (many) and *meros* (parts). The word "plastic" comes from the Greek word *plastikos*, meaning "able to be moulded."

Nylon was first used as toothbrush bristles, and Teflon was listed in the Guinness Book of World Records as the most slippery substance in the world. DuPont described nylon as being made from coal, water, and air. Nylon can be made into fine, shiny threads that are as strong as steel and yet more elastic than common natural fibres. Besides its use in women's stockings, nylon was used for parachutes, tents, and aircraft tire cord during World War II. Today, nylon is used in many items such as computer circuit boards, fishing lines, surgical thread, clothes, carpets, fire hoses, and ropes. Although Teflon is commonly associated with cookware, it is largely used today as insulation for wires and cables.

Since plastics cannot be decomposed easily, plastic wastes remain in the environment. Julian Hill, the only surviving member of the team that developed nylon, is worried about the growing threat of plastic products to the environment. "I don't think it is all good," he says, "I think the human race is in danger of being smothered in its own plastics."

What do you think?

Fig. 10 *Nylon is strong enough to make cables, yet soft enough to make women's stockings.*

Fig. 11 *Teflon is best known for its use on non-stick cookware.*

"Mercury(II)" in the name means the compound contains Hg^{2+} ions.

Minamata disease is a nerve disorder caused by high levels of mercury in the body. It was first discovered in Japan, where industries were dumping large amounts of mercury into the sea. The mercury taken into the body attacks all the tissues, including the brain. The effects are irreversible.

Mercury

The earth dug up for the James Bay project contains the ore known as cinnabar, which is mercury(II) sulfide. The unearthed cinnabar dissolves in the waters of the reservoir at James Bay, and bacteria convert the mercury salt into a compound called methylmercury. As fish eat the tiny organisms that form methylmercury, the level of mercury in the fish builds up to quite a high level. Mercury builds up to an even higher level in the people who eat the fish. The Crees of northern Québec have shown symptoms of Minamata disease.

At its research laboratory in Varennes, Hydro-Québec monitors the concentration of mercury in the waters of the James Bay reservoir.

CHEMICAL TIDBITS

DOES AN APPLE A DAY KEEP THE DOCTOR AWAY?

In early 1989, an environmental organization in the United States claimed that a by-product of Alar present in apples causes cancer in animals. Alar is the commercial name of an organic acid, daminozide. It is used to promote the ripening of apples on the tree. Because Alar was detected not only on the skin but also in the pulp of the apple, there was considerable alarm. It was feared that children, who consume many apples and apple products, would be especially harmed by Alar.

While members of the industrial chemistry community insisted there was no immediate risk from the consumption of apples sprayed with Alar, they could not assure the public that there would be no long-term effects. Consumers and governments exerted pressure on the companies that produced Alar, and sales of the chemical were halted.

Scientists and government regulators are used to relying on test data to determine the level of risk involved in using a particular chemical. Minimal risk levels are considered acceptable for substances that provide important benefits to society. The fear is that drastic reductions in our food supply, already not enough to feed the world, would result if chemicals such as Alar and certain pesticides were not used. Many members of the public, however, are concerned about unfamiliar chemicals imposed by profit-making companies. Some citizens are demanding alternatives with no risk to future generations, and the chemical industries are beginning to respond.

Fig. 12 *What do you think about the use of chemicals in farming?*

PCPs and PCBs

Fig. 13 *Wooden utility poles can be preserved with PCPs.*

About 95% of the utility poles in the province of Québec have been treated with pesticides known as PCPs, polychlorophenols. PCPs keep the wood from being eaten away by insects and micro-organisms (Figure 13). In 1987, the province used 1400 t of these pesticides. Hydro-Québec is studying how PCPs enter the environment and how long they remain there. It is also seeking a replacement that is less environmentally harmful, although at this point it considers the level of damage due to PCPs to be small compared to their benefits.

Ever since the fire at St-Basile-le-Grand in 1988 (Figure 14), public attention has been called to another group of chemicals known as PCBs, or polychlorinated biphenyls. Stockpiles of PCBs exist in various places. Canada's largest PCB dump is in Shawinigan, Québec.

PCBs are thick, oily liquids that do not dissolve in water and do not conduct electricity. They are fire-resistant, but they easily vaporize at temperatures over 40°C. PCBs have been used as coolants and insulators for electrical capacitors and transformers. When it was determined that PCBs could be harmful to the environment and to human health, their manufacture was stopped in 1977. However, around 40 000 000 kg had already been imported into Canada. Once they are released into the environment, they are incorporated into aquatic plants, animals, and finally into human beings. Winds have carried PCB vapours all over the world.

Fig. 14

A **capacitor** is a device used to store electricity.

Although no direct link has been established between PCBs and cancer in human beings, they have been shown to cause skin and eye irritation, headaches, vomiting and fever (Figure 15). One of the problems facing our society is the safe disposal of the PCBs now stored in stockpiles of used transformers and capacitors.

Fig. 15 *PCBs (polychlorinated biphenyls) are dangerous organic compounds that must be disposed of properly.*

The Greenhouse Effect

Have you ever sat in a car on a hot, sunny day? As the sunlight pours in through the windows, it heats up the inside of the car. In turn, the car radiates back some, but not all, of the heat. Some of the heat is "trapped" inside the car. This trapping of heat is called the **greenhouse effect**. Our earth is surrounded by a layer of atmospheric gases that traps some of the heat from the sun. It is feared that, as a result of pollution, the changes in the earth's atmosphere will cause the temperature of the earth to rise gradually (Figure 16). Some scientists predict that the earth's temperature will increase by 2°C to 6°C, resulting in great climatic changes, such as the creation of more desert and the melting of the polar ice caps. The principal greenhouse gas is thought to be carbon dioxide.

It is estimated that 4 500 000 000 to 5 500 000 000 t of carbon in the form of carbon dioxide is released into the atmosphere each year. That is 25% more than the annual rate 100 years ago. The rate may double from today's value by the middle of the next century, if measures are not taken to reverse the trend. Although

Canada produces only a bit more than 2% of the world's carbon dioxide, its citizens have the highest per capita production: 4 t for each person per year!

Scientists believe there are only two effective means to reverse the greenhouse effect — to burn less fossil fuel and to plant more trees.

Fossil fuels include oil, natural gas, gasoline, and coal. They are called fossil fuels because they were formed by the decay of once living matter from the age of the dinosaurs.

Fig. 16 *Air pollution may cause the temperature of the earth to increase.*

Ozone

Ozone is a form of oxygen with three atoms per molecule: O_3. Ozone has a sharp and fresh smell, which you may detect if you stand by a heavily used photocopier.

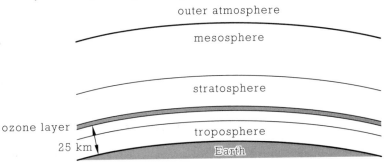

Fig. 17 *The Earth's ozone layer.*

At an international meeting held in Montréal in 1987, the representatives of 24 countries agreed to save the ozone layer that protects the earth from the sun's harmful ultraviolet rays (Figure 17).

The Montréal Protocol called for a reduction in the use of chemicals known as CFCs, which react with the ozone in the atmosphere and make holes in the protective ozone layer.

Repeated exposure to ultraviolet radiation increases the occurrence of skin cancer.

387

CFCs or chlorofluorocarbons are volatile gases used in plastics manufacture, refrigerators, and air conditioners. Their use as propellants in aerosols has already been drastically reduced in response to the possible danger they pose to the environment. Canada, the U.S., and the European Economic Community plan a total ban of CFCs by the turn of the century.

While the lack of ozone high in the atmosphere is a problem, too much ozone at ground level is hazardous. Ozone is formed when nitrogen oxides in the fumes from traffic and industry react with oxygen in the presence of sunlight. Ground-level ozone can affect the functioning of the lungs, and can cause breathing problems and coughing. Environment Canada reported in 1990 that it could cost Canadian consumers more than $800 million annually by the year 2005 to reduce ground-level ozone to a safer level by controlling emissions.

Questions

1. (a) What role do phosphates play in detergents?
   (b) Why are stores offering their customers "phosphate-free" detergents?

2. (a) What kind of packaging material is not biodegradable?
   (b) What can you do about discouraging the use of excess packaging?

3. (a) Which gas is referred to as the greenhouse gas?
   (b) Consult research sources to determine the other gases involved in the greenhouse effect.
   (c) How can you reduce the level of greenhouse gases in our environment?

4. (a) Why is ozone important for our health?
   (b) Can ozone also damage our health? Explain.
   (c) What can we do to repair the ozone layer that protects the earth?

5. Progress has been made in reducing some forms of air pollution. As an example, research and report on the changes in airborne lead pollution since the 1970s. Has the progress in reducing lead pollution been worldwide?

13.6 Handling Our Chemical Waste

The Stablex plant in Blainville, just 25 km north of Montréal, is licensed by the Québec government to treat 100 000 t of toxic industrial waste a year. Toxic wastes that are not recycled, such as acids, cyanides, and sodium hydroxide, are neutralized and then converted into concrete-like blocks for storage in its landfill site. In 1987, the Stablex plant reported operating at only 70% of capacity. Most of the toxic waste it treated was from the United States — only 14 000 t came from the province's industries in the first 10 months of 1987. Many Québec companies are believed to be dumping their waste into rivers and sewage systems. Thousands of Québec companies do not even report how much toxic waste they produce and how they dispose of it, contrary to the requirements of provincial law.

Organic waste, such as the oils, solvents, and liquid cleaners used in the paint and printing industries, makes up half of the industrial waste in the province. These materials can be recycled, sent to treatment centres outside Québec, or burned in the Laidlaw Environmental Services incinerator in Mercier. Laidlaw burns about 40 000 t of used oils a year, mostly from Québec.

It is easy to blame industries for all our problems, but to what extent do individual members of the public create chemical waste?

Southam News reports that nearly 300 000 000 L of motor oil is heedlessly dumped into Canadian sewers each year by Canadian do-it-yourself mechanics. This volume amounts to almost eight times the volume of oil spilled off the coast of Alaska by the Exxon Valdez in 1989.

According to Environment Canada, 35% of the mercury released into the Canadian environment comes from used batteries, such as those used in radios, clocks, and flashlights (Figure 18).

Fig. 18

How much waste do you think you release into our environment in a day? a week? a year? How could you measure the quantity of waste of various kinds you do produce? Get together in small groups and discuss these questions. When you are ready, share your ideas with the class.

Can you make a reliable estimate of the quantities of the various kinds of waste that you produce, or is it necessary to monitor your activities over time? If you feel that such monitoring is necessary, devise and discuss a reliable method for measuring and classifying the waste you produce. Carry out the agreed procedure and discuss your findings with the class. Do your findings agree with your estimates of the quantities of waste of various types that you produce?

Questions

1. (a) Suggest ways you could reduce the quantity of chemical waste you release into the environment.
   (b) How do your suggestions compare with methods currently used in society?

2. (a) What sort of methods are used in your community to collect and dispose of toxic chemicals, like paint thinners, cleaning solvents, and lawn care chemicals?
   (b) What improvements would you suggest to make the collection and recycling of toxic chemicals easier?

3. How would you reduce the quantity of non-recyclable toxic chemicals in your community?

4. In May 1991, the environmental group Société pour vaincre la pollution (SVP) claimed that previous owners of the toxic waste incinerator in Mercier had buried as many as 600 barrels of toxic waste on the site, instead of destroying the waste. It is possible that the environmental damage could be many times worse than that from the PCB fire in St-Basile-le-Grand. Who do you think should be responsible for cleaning up the site and for any environmental damage it causes? the previous owners? the present owner? some level(s) of government? How much environmental damage has been documented and what are the future possibilities? Is the situation now under control? Research the latest information and prepare a report.

13.7 The Beginnings of the Chemical Industry in Québec

Go through the Yellow Pages of your local telephone directory and list all the industries whose work involves chemicals. You may be surprised to discover how few industries do not have a "chemical connection."

Group the manufacturing industries according to the type of product they manufacture. In your area, what is the most manufactured product? Is it used within the province, sold elsewhere in Canada, or sold to the United States and other countries? Where is its biggest market? Is this product manufactured in other parts of Québec? Why do you think the company chose your area to establish itself?

How does the manufacture of that product affect your environment? How many people are employed in manufacturing this product?

From its earliest days, Québec has utilized its immense natural resources and abundant hydro-electric power to build its industrial base. Even in 1743, the forges at St Maurice maintained a production level of 175 000 t of iron a year. An industrial sector comprising textiles, shoe manufacture, railway goods, food, clothing, and wood and wood-related materials was thriving by 1860.

The lure of rich natural resources and the availability of 40% of the world's unharnessed and conveniently located sources of hydro-electric power brought substantial foreign investment. This came principally from the United Kingdom and the United States, in the areas of hydro-electric development, the establishment of pulp and paper mills, and metallurgy — especially the production of aluminum. Québec was a major innovator in the use of electrolytic methods to produce chemicals at the turn of the century.

In 1904, Thomas Willson brought his expertise in building successful electric furnaces from Niagara to La Mauricie where he established Shawinigan Incorporated. The company produced calcium carbide, CaC_2, from lime, CaO, and coke, a form of carbon.

The beginning of World War I demanded greater production of acetylene from calcium carbide. Shawinigan became the chemical centre of Canada as it expanded its production to include other industrial chemicals related to acetylene, that is, acetaldehyde, acetic acid, and acetone. Over the next few generations, Shawinigan also produced vinyl resins and plasticizers, becoming a key player in the munitions manufacture for Canada during World War II.

Vinyl resins and plasticizers are used in the manufacture of plastic products.

CHEMICAL TIDBITS

A CHANCE DISCOVERY

The first person to produce calcium carbide cheaply and on a large scale was a Canadian called ''Carbide'' Willson. He discovered his process by accident.

Thomas Leopold Willson was born in Princeton, near Woodstock, in Upper Canada in 1860, and was educated at Hamilton Collegiate Institute. As a young man he started experimenting in the basement of his home. He made all kinds of gadgets and a number of chemicals. With a blacksmith friend, he built one of the first electric dynamos in Canada. In 1880, he used the dynamo to run the first arc light in Hamilton.

He became an electrical engineer and went to work in Spray, North Carolina. Here in 1892, in the Willson Aluminum Company, Willson tried to prepare calcium. He heated a mixture of lime (calcium oxide) and coal-tar in a high temperature electric furnace, and obtained a dark molten mass. On cooling, the molten mass became a highly crystalline, brittle solid. He threw some of the solid into a bucket of water. Calcium would have reacted with the water to form hydrogen gas. Instead of hydrogen, the reaction with water produced another gas, which started to burn because the top of the solid was still glowing. Unlike hydrogen, the gas burned with a smoky flame and gave off large quantities of black soot. The gas was acetylene, and the solid he had made was calcium carbide.

$$\text{calcium carbide} + \text{water} \rightarrow \text{calcium hydroxide} + \text{acetylene}$$
$$CaC_2 + 2\,H_2O \rightarrow Ca(OH)_2 + C_2H_2$$

And so Thomas Willson was responsible for developing the commercial calcium carbide process for the manufacture of acetylene. He sold his American patents to the Union Carbide Company and moved to Merritton, Ontario, where he organized the Willson Carbide and Acetylene Works. In 1911, he sold all his Canadian manufacturing rights to the Canada Carbide Company.

Fig. 19 *"Carbide" Willson was the first person to produce acetylene cheaply.*

Fig. 20 *Refining petroleum results in very important products.*

Following World War II, the petroleum refining industry became established at the east end of Montréal island (Figure 20). Shawinigan Carbide transferred its organic chemistry to Varennes on the south shore of the St. Lawrence River. While not among one of the largest producers of petroleum-related chemicals, Québec enjoyed a competitive position on world markets until a rise in oil prices and the utilization of natural gas reserves in Western Canada reduced Québec's economic position. In an attempt to recover in the 1970s, the Société générale de financement established Ethylec Incorporated, which ran two petroleum plants in the Montréal area in conjunction with the Petromont consortium. PetroCanada began to produce certain so-called aromatic hydrocarbons, such as benzene, toluene, and xylene, with its takeover of the Petrofina plants near Montréal.

13.8 Some Modern Chemical Industries

Aluminum

Aluminum at one time was so valuable that a block of it was presented to Queen Victoria at her coronation. Thanks to the economical method of aluminum production proposed independently by Charles Hall of the United States and Paul Louis Toussaint Heroult of France, the price of aluminum fell drastically.

Fig. 21 *Many products are made from aluminum, including a variety of food packages.*

The ore bauxite is 50% to 60% aluminum oxide, also known as alumina. The aluminum oxide is separated from the bauxite and dissolved in a molten electrolyte. When the electricity is turned on, the aluminum is plated onto a carbon-lined cell, which serves as the cathode. Oxygen gas collects around the anode.

Because the province is rich in electrical power, Québec was able to step into the forefront of the world's aluminum production. In 1985, Québec was responsible for the production of 8.3% of world supply of aluminum and aluminum products from the six smelters maintained by Alcan Aluminum Limited (Figure 21). Canada consumes only 25% of that production. The rest is exported, mainly to the United States.

Other Minerals

Nonferrous metals, such as copper, zinc, nickel, and lead, are refined from sulfur-containing ores and turned into various products.

In 1986, Québec was responsible for the following percentages of the national production of minerals.

zinc	4.0%
copper	8.7%
iron	36.8%
gold	28.2%
silver	4.1%
asbestos	80.5%
titanium and niobium	100%

The mining of uranium is expected to play a role in the future economy of Québec.

Noranda Mines Limited is one of the largest copper refiners in the world. Noranda developed a simple furnace called the Noranda reactor, which provides a continuous production of copper. Copper ore enters one end of a horizontal cylinder, where it is made into pure copper by the action of oxygen-enriched air:

$$CuS + O_2 \rightarrow Cu + SO_2$$

The copper is removed from one side, while the slag, which contains some leftover copper, is taken for further processing and recycling.

Pulp and Paper

Québec's 53 plants produce 35% of the Canadian output in newsprint, making the province one of the top ten of the world's producers of newsprint. Timber, wood pulp, and newsprint constitute 20% of Québec's exports. One-third of these wood products goes to the United States.

Pharmaceuticals

The pharmaceutical industry was established in Canada in 1879 when E. B. Shuttleworth founded a pharmaceutical firm in Toronto. Most of the pharmaceutical industry in Canada is owned by foreign firms. Connaught Laboratories in Toronto and Institut Armand-Frappier in Montréal produce biological materials such as vaccines, insulin, and blood products. The Canadian pharmaceutical industry is the ninth largest in the world, and nearly half of those employed in the pharmaceutical industry work in the province of Québec. Approximately 25 new drugs are marketed each year and 40% of those improve on currently used drugs.

In the 1960s, Merck Frosst Limited in Montréal started work on a new drug to control hypertension or high blood pressure. Their study and research in a new drug, timolol maleate. The drug underwent testing by the federal drug regulatory agencies and went on the market in the 1970s. It is considered to be quite effective in a variety of medical procedures.

CHEMICAL TIDBITS

ARMAND FRAPPIER

Armand Frappier (Figure 22), chief of laboratories at Hôpital Saint-Luc and a professor of bacteriology at the University of Montréal, founded what was later to be named the Institut Armand-Frappier in 1938. The institute is a teaching and research facility which provides some health services and produces some biological products for use in industry. Dr. Frappier prepared a vaccine to fight tuberculosis and, with the help of his colleague Paul Lemonde and his daughter Lise Davignon, showed this vaccine is useful in preventing cases of infant leukemia. Dr. Frappier studied how infections take hold and how they are resisted, and he worked to establish one of the few laboratories in the world devoted to the study of Hansen's disease or leprosy.

Fig. 22 *Armand Frappier.*

Questions

1. What are the major resources of the province of Québec?

2. What effect has abundant and inexpensive hydro-electric power had on the growth of chemical industries in Québec?

3. How has the Hall/Toussaint process played a role in the growth of the aluminum industry in the province?

13.9 Recycling Our Wastes

Fig. 23 *Around Montréal and Québec City, there are five large dumps where more than 11 million tires are piled up. Fires, like the one in St-Amable, could add appreciably to the level of carbon dioxide and PCBs in the atmosphere.*

It has been estimated that Quebeckers discard 6 500 000 t of waste each year. That amounts to 240 green garbage bags per family! The total production of garbage seems to be growing at a rate of as much as 7% a year. More than three-quarters of the waste ends up in sanitary landfills. The rest is incinerated, and only 10% is recycled.

According to studies done in the province of Québec in 1989, the contents of an average home garbage pail were the following:

Material	Québec
paper and cardboard	41.7%
organic matter	28.2%
other material	7.9%
glass	6.2%
plastic	4.2%
ferrous metals	5.7%
textiles	5.6%

Nearly all these items are recyclable. An American study shows we can save 80% on energy and cut 77% of post-consumer solid waste if we reuse a corrugated container five times. The recycling of steel uses 74% less energy than manufacturing more steel. The saving rises to 95% in the case of aluminum.

Recycling helps us in four important ways:

1. Because it takes less energy to recycle goods than to manufacture them from raw materials, we save on energy.

2. We save on non-renewable resources like aluminum, iron, and copper. We also reduce the quantities of renewable resources, such as lumber, that we have been destroying faster than we have been replacing them.

3. Reducing the quantity of packaging saves money for us at the cash register.

4. By lowering the quantity of garbage we send to the landfills and reducing the quantity of goods produced, we add less pollution to the environment.

In some neighbourhoods there are places to drop off newspapers, glass, or plastic for recycling. To encourage the co-operation of the public, a few communities have asked their citizens to sort their garbage and place it in special "blue boxes" for collection from their homes. The sorted garbage is then sold to recycling plants. The large pulp and paper industry in Québec has already begun extending its work into the field of recycling.

Biodegradable plastics, which have some starch mixed in, may not be such an improvement. They break up into tiny pieces, which are more likely to be transported by the wind or water than the usual plastic materials. What is more, biodegradable plastics cannot be recycled.

CHEMICAL TIDBITS

RECYCLING PLASTICS

While incinerating plastics could provide heat energy, we pay a price for sending reusable material up the smokestack. Eighty-seven percent of all plastics sold are recyclable **thermoplastics**. The term "thermoplastic" is applied to a plastic that can be remelted and reformed into other plastic products with little or no lowering of quality. Thermoplastics include the widely used high- and low-density polyethylene. The high-density polyethylene (HDPE) is used for rigid containers, such as milk and water jugs, household product containers, and motor oil bottles. The low-density polyethylene (LDPE) is used for packaging film and shopping bags. PET or polyethylene terephthalate, another thermoplastic, is used to make bottles for carbonated beverages. Polystyrene (PS) is used in the form of a foam to make food containers, drinking cups, and trays. In its rigid form, it is used to make plastic cutlery. PVC or polyvinylchloride can be used in plumbing or construction, in some consumer packaging, vinyl flooring, car upholstery, rain boots, and rain coats (Figure 24). All of these plastics, as well as others, can be reclaimed and reprocessed to save energy and natural resources.

The technology of plastics recycling is not new. Plastic manufacturers already recycle the scrap plastic leftover in manufacturing. All that prevents the recycling of plastic on a larger scale is the difficulty of getting used plastics back from the consumer.

Dow Canada, in co-operation with Domtar Inc. of Montréal, has built a PET and high density polyethylene recycling plant designed to treat about 400 million plastic bottles and containers a year.

Recycling is one of the three "R's" for the protection of the environment from waste pollution. The other two are Reduction and Reuse. Reduction means we avoid buying products that are over-packaged and we buy only what we know we will use. Reusing is repairing rather than replacing broken items and finding other uses for empty containers. What can you do in your home to make our environment a more healthy and pleasant place and to conserve our resources for the future?

Fig. 24 *PVC can be recycled.*

CHEMICAL TIDBITS

CAN A CHEMICAL COMPANY BE A GOOD NEIGHBOUR?

Rohm and Haas Canada Inc. has environmental control programs designed to reduce or eliminate waste and harmful plant emissions. Air pollution is reduced when exhaust gases are cleaned by passage through scrubbing devices. Water pollution and waste are reduced or eliminated through recycling or improved methods of production. Incineration is used for any remaining waste disposal. A key part of environmental control is regular testing and analysis of air and groundwater samples.

Chemical plants such as Rohm and Haas Canada Inc. produce fertilizers, pesticides, and synthetic polymers used to make common products.

Questions

1. Identify companies in your region that recycle consumer goods. What materials do they recycle?

2. Consider the three "R's" for the protection of the environment.
   (a) Do you think of all three as equally important and effective? If so, justify your answer. If not, list them in decreasing order of importance and justify your answer.
   (b) Compare your answer to (a) with the views of environmental groups.

3. Describe the effects that society's increased emphasis on recycling has on each of the following:
   (a) your daily life
   (b) life in your province

4. Organize a debate of the following statement, with one side speaking for it, the other against: The government should have no say in the materials used to manufacture consumer goods.

5. Identify at least three examples of non-polluting chemicals that you could substitute for polluting chemicals in everyday use.

6. Do you believe that existing means of disposal and recycling of chemical wastes are adequate in your region? Support your answer with specific examples.

POINTS · TO · RECALL

- Combustion is the rapid reaction of a fuel with oxygen to give off heat and light.
- The oxide of a nonmetal forms an acid in water; the oxide of a metal forms a base in water.
- Acid precipitation results from nonmetal oxides in the air. They dissolve in the water that makes rain, snow, and mist.
- The principal sources of acid precipitation are the burning of fossil fuels, like coal and petroleum products, and the refining of ores.
- Phosphates in detergents cause an increase in the growth of algae in lakes. The algae consume the oxygen that animals need to live, and so the lakes "die."
- Biodegradable substances are broken down into smaller substances by bacteria in the environment.

- Mercury is washed into the water system by acid rain. Mercury accumulates in the food chain, possibly producing Minamata disease in people who eat the fish from contaminated lakes.
- PCPs are pesticides used to preserve utility poles. Some people believe that PCPs harm the environment.
- PCBs are nonelectrolytic oils that have been used in transformers. One of the major problems facing environmentalists is the disposal of stockpiled PCBs. They cannot be burned in the usual manner because their fumes could make people ill.
- The greenhouse effect is a gradual warming of the earth because of heat trapped by certain atmospheric gases, especially carbon dioxide.

- A reduction in the burning of fossil fuels and the planting of trees would help diminish the greenhouse effect.
- Ozone is a form of oxygen gas with three atoms per molecule. In the upper atmosphere it serves to filter out harmful radiation from the sun.
- CFCs in propellants and refrigerants are creating holes in the protective ozone layer. Their replacement in household products would help in restoring the ozone layer.
- Organic wastes, such as solvents and oils, can be recycled, treated, or burned at licensed facilities. Non-organic wastes, such as acids, cyanides, and sodium hydroxide, are treated and stored in landfills.

- The wealth of mineral resources and hydro-electric power has made Québec a centre for the chemical industry in Canada.
- Important chemical industries in Québec include the aluminum industry, the refining of ores, pulp and paper, and pharmaceuticals.
- Recycling our wastes is beneficial because it takes less energy to recycle goods than to manufacture new ones. Also, non-renewable resources are saved, and renewable resources last longer.
- Less packaging means less garbage, and less garbage means less pollution.
- Thermoplastics can be recycled with negligible loss of quality.

REVIEW · QUESTIONS

1. How does the combustion of fuels affect our environment?
2. What is acid precipitation and what causes it?
3. What features of the geology of your province make it easily damaged by acid rain?
4. Complete and balance the following equations:
   (a) $C + O_2 \rightarrow$
   (b) $S + O_2 \rightarrow$
   (c)   + $\rightarrow MgO$
   (d)   + $\rightarrow P_2O_5$
   (e)   + $\rightarrow N_2O_5$
   (f) $SO_3 + H_2O \rightarrow$
   (g) $N_2O_5 + H_2O \rightarrow$
   (h) $P_2O_5 + H_2O \rightarrow$

5. What role do the rain forests play in preserving our environment?
6. (a) How do mercury, PCPs, PCBs, and CFCs enter our environment?
   (b) How can mercury, PCPs, PCBs, and CFCs harm our environment?
   (c) What can you do to bring about a reduction in the quantities of mercury, PCPs, PCBs, and CFCs in our environment?
7. (a) Briefly explain the greenhouse effect.
   (b) What gases are responsible for the greenhouse effect?

8. What is ozone and why is the ozone layer around the earth important to our environment?
9. What substances found in your home are classified as recyclable organic wastes and which are non-organic wastes that cannot be recycled?
10. What are the principal chemical industries found in:
    (a) Québec?
    (b) your vicinity?
11. How are the principal chemicals produced in Québec used?
12. What percentage of our wastes is recyclable?
13. How can recycling preserve our renewable and non-renewable resources?
14. What recycling programs are in use
    (a) in your community?
    (b) in your school?
15. One of the greatest roadblocks to a greener and cleaner environment is the lack of public concern. How can you encourage your family and your neighbours to take an active role in recycling?

Key words are indicated by boldface references; (c) refers to a chart; (f) refers to a figure.

Appendix 1

Precision and Accuracy

Precision describes the reproducibility of measurements and results, whereas accuracy describes how close they are to the correct values. Precise values may or may not be accurate. Suppose that you determine the mass of an object to be 4.8514 g, 4.8515 g, and 4.8513 g in three trials. These values are very close together. In other words, they are highly reproducible or very precise. If the correct mass is 4.8514 g, then the three mass determinations are also very accurate. But suppose that the balance is incorrectly adjusted, so that all mass readings are too high by 1 g. Then, no matter how precise the readings are, they are all inaccurate.

An analogy often used to explain the difference between precision and accuracy is to describe an archer shooting at a target. If all the arrows are close together on the target, the archer's aim is very precise. If the arrows are bunched together in the centre of the target, the archer's aim is also very accurate. If the arrows are close together but not near the centre, the archer's aim is precise but not accurate. If they are spread all the way around the outside of the target, the archer's aim is neither precise nor accurate.

Significant figures (sig. figs.) express the precision (not the accuracy) of measurements and calculated results.

In a correctly recorded measurement, the figures you are sure of, plus the first estimated or uncertain figure, are all significant figures. For example, if you were determining the mass of an object on a triple beam balance, you might find a value of between 4.1 g and 4.2 g. To arrive at a more precise value for the mass, you must estimate the next figure. If the mass seems to be closer to 4.2 g than to 4.1 g, you might estimate it at 4.17 g. The "7" is uncertain; another student might estimate it as a 6 or an 8, and you might change your estimate in a repeat trial. However, the uncertain figure must be recorded as part of the measurement. To omit the last figure because it is estimated is to record a less precise value than the instrument warrants.

In general, when you use a measuring instrument, you should estimate the final figure of the measurement between the smallest divisions of the scale on the instrument. For example, if a triple beam balance has 0.1 g divisions as the smallest divisions on the scale, you should expect to estimate masses on this balance to the nearest 0.01 g. Even if a mass value seems to exactly coincide with a division, you should express the value to 0.01 g precision. If the mass seems to be exactly 5.8 g, for example, you should record the mass as 5.80 g, if the smallest divisions on the scale are 0.1 g apart.

When you see a measurement that someone else has recorded, how do you know the number of significant figures in it? Here are some useful rules:

1. All non-zero digits are significant.
   Example: In the measurement 4.17 g, described above, there are three sig. figs.

2. Zeros may or may not be significant.
   (a) Zeros between non-zero digits are significant.
   Example: A length of 1.008 m has four sig. figs.
   (b) Zeros that precede the first non-zero digit are never significant.
   Example: A volume of 0.00015 mL has two sig. figs.
   (c) Zeros that follow the last non-zero digit are only significant if they arise from the correct use of a measuring instrument.

   Example 1: A measured length of 0.8400 cm has four sig. figs. The final two zeros in this number are only recorded if they reflect the precision of the instrument. In this case, the smallest divisions on the instrument are 0.001 cm apart, so measurements must be recorded to the nearest 0.0001 cm.

   Example 2: A mass of 3200 kg may have two, three, or four sig. figs. You cannot tell from looking at the number, because the zeros before the decimal point must be there to convey the size of the number. (You cannot, for example, write 3200 kg to two sig. figs. as 32 kg! If you do, you change the number by a factor of 100.)

The ambiguity can be avoided through the use of scientific notation. Thus, we can write 3200 kg to two sig. figs. as 3.2×10^3 kg, to three sig. figs. as 3.20×10^3 kg, and to four sig. figs. as 3.200×10^3 kg. In this book, assume that the zeros recorded in such an example *are* significant. In other words, assume that a mass of 3200 kg has four sig. figs. (Such a mass measurement comes from an instrument with its smallest divisions 10 kg apart.)

When you carry out calculations that involve measurements, follow these rules:

1. In a multiplication or division, the answer has the same number of significant figures as the measurement with the smallest number of significant figures.

   Example: $2.16 \times 3.401 = 7.34616$
   $$= 7.35 \text{ (to three sig. figs.)}$$

2. In an addition or subtraction, the uncertainty in the answer is in the same column as the earliest uncertainty in any of the measurements being added or subtracted.

   Example: 3.623
   $$\underline{-\,3.01}$$
   $$= 0.613$$
   $$= 0.61 \text{ (to two sig. figs.)}$$
   Note that the earliest uncertainty in any measurement comes in the one-hundredths column, so our answer must end in the one-hundredths column.

Appendix 2

The Equation and Slope of a Straight Line

The equation of a straight line is of the general form $y = mx + c$, where y is the variable on the vertical axis, x is the variable on the horizontal axis, m is the slope of the line, and c is the y-intercept. The y-intercept is the value of y at which the line cuts the vertical axis. If the line passes through the origin, then $c = 0$, and the equation of the line simplifies to $y = mx$.

Suppose we determine the values of two variables, x and y, in an experiment. Then we plot y against x and obtain a straight line. How can we determine the equation of the line? Obviously, we need the values of m and c. If m is found to be 2 and c is 3, for example, then the equation is $y = 2x + 3$.

You may recall that the slope of a straight line is defined by the equation: slope $= \dfrac{\text{rise}}{\text{run}}$

To determine the slope, take any two points on the line and divide the difference in their y-values by the difference in their x-values. In Figure 1,

$$\text{slope} = \frac{y_2 - y_1}{x_2 - x_1}$$
$$= \frac{2 - (-2)}{3 - (-1)}$$
$$= \frac{4}{4}$$
$$= 1$$

As you can see from Figure 1, the line crosses the y-axis at $y = -1$. Therefore, $c = -1$. Thus, the equation of the straight line is $y = 1x - 1$, which we usually write as $y = x - 1$.

Figure 1

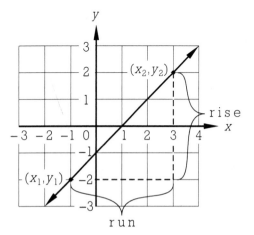

Symbols and Valence Values of Some Elements and Radicals

Element	Symbol	Valence Value
aluminum	Al	3
antimony	Sb	3, 5
arsenic	As	3, 5
barium	Ba	2
beryllium	Be	2
bismuth	Bi	3, 5
boron	B	3
bromine	Br	1
cadmium	Cd	2
calcium	Ca	2
carbon	C	4
cesium	Cs	1
chlorine	Cl	1
chromium	Cr	2, 3
cobalt	Co	2, 3
copper	Cu	1, 2
fluorine	F	1
gold	Au	1, 3
hydrogen	H	1
iodine	I	1
iron	Fe	2, 3
lead	Pb	2, 4
lithium	Li	1
magnesium	Mg	2
manganese	Mn	2, 3
mercury	Hg	1, 2
nickel	Ni	2, 3
nitrogen	N	3
oxygen	O	2
phosphorus	P	3, 5
potassium	K	1
silicon	Si	4
silver	Ag	1
sodium	Na	1
strontium	Sr	2
sulfur	S	2
tin	Sn	2, 4
zinc	Zn	2

Radical	Symbol	Valence Value
acetate	$C_2H_3O_2$	1
ammonium	NH_4	1
arsenate	AsO_4	3
arsenite	AsO_3	3
bromate	BrO_3	1
carbonate	CO_3	2
chlorate	ClO_3	1
chlorite	ClO_2	1
chromate	CrO_4	2
cyanate	CNO	1
cyanide	CN	1
dichromate	Cr_2O_7	2
dihydrogen phosphate	H_2PO_4	1
hydrogen carbonate	HCO_3	1
hydrogen phosphate	HPO_4	2
hydrogen sulfate	HSO_4	1
hydrogen sulfide	HS	1
hydrogen sulfite	HSO_3	1
hydroxide	OH	1
hypochlorite	ClO	1
iodate	IO_3	1
monohydrogen phosphate	HPO_4	2
nitrate	NO_3	1
nitrite	NO_2	1
perchlorate	ClO_4	1
phosphate	PO_4	3
phosphite	PO_3	3
sulfate	SO_4	2
sulfite	SO_3	2

The Periodic Table of the Elements

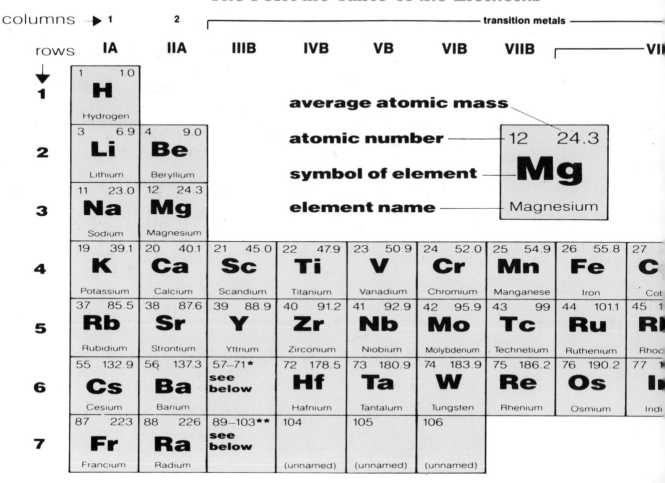

columns → 1   2   transition metals

rows   IA   IIA   IIIB   IVB   VB   VIB   VIIB   VII

1	1.0 **H** Hydrogen							

average atomic mass
atomic number — 12   24.3
symbol of element — **Mg**
element name — Magnesium

Row 2: 3 6.9 **Li** Lithium | 4 9.0 **Be** Beryllium

Row 3: 11 23.0 **Na** Sodium | 12 24.3 **Mg** Magnesium

Row 4: 19 39.1 **K** Potassium | 20 40.1 **Ca** Calcium | 21 45.0 **Sc** Scandium | 22 47.9 **Ti** Titanium | 23 50.9 **V** Vanadium | 24 52.0 **Cr** Chromium | 25 54.9 **Mn** Manganese | 26 55.8 **Fe** Iron | 27 **C** Cob

Row 5: 37 85.5 **Rb** Rubidium | 38 87.6 **Sr** Strontium | 39 88.9 **Y** Yttrium | 40 91.2 **Zr** Zirconium | 41 92.9 **Nb** Niobium | 42 95.9 **Mo** Molybdenum | 43 99 **Tc** Technetium | 44 101.1 **Ru** Ruthenium | 45 1 **Rt** Rhod

Row 6: 55 132.9 **Cs** Cesium | 56 137.3 **Ba** Barium | 57–71* see below | 72 178.5 **Hf** Hafnium | 73 180.9 **Ta** Tantalum | 74 183.9 **W** Tungsten | 75 186.2 **Re** Rhenium | 76 190.2 **Os** Osmium | 77 **Ir** Iridi

Row 7: 87 223 **Fr** Francium | 88 226 **Ra** Radium | 89–103** see below | 104 (unnamed) | 105 (unnamed) | 106 (unnamed)

*Lanthanide series

57 138.9 **La** Lanthanum	58 140.1 **Ce** Cerium	59 140.9 **Pr** Praseodymium	60 144.2 **Nd** Neodymium	61 147 **Pm** Promethium	62 150.4 **Sm** Samarium	63 **E** Europ

**Actinide series

89 227 **Ac** Actinium	90 232.0 **Th** Thorium	91 231 **Pa** Protactinium	92 238.0 **U** Uranium	93 237 **Np** Neptunium	94 242 **Pu** Plutonium	95 **A** Ameri

There are different ways for numbering the groups in the periodic table.

In this book, the eight longer columns are numbered as Groups 1 to 8, and the elements between Groups 2 and 3 are called transition metals.